WHALES AND DOLPHINS OF AOTEAROA NEW ZEALAND

WHALES

AND DOLPHINS OF AOTEAROA NEW ZEALAND **BARBARA TODD**

TE PAPA

PRESS

First published in New Zealand in 2014 by
Te Papa Press, PO Box 467, Wellington, New Zealand
tepapapress.co.nz

Developed partly in conjunction with the exhibition
Whales | Tohorā, developed by the Museum of New Zealand
Te Papa Tongarewa, Wellinton, New Zealand.
tepapa.govt.nz

TE PAPA® is the trademark of the Museum of New Zealand
Te Papa Tongarewa
Te Papa Press is an imprint of the Museum of New Zealand
Te Papa Tongarewa

A catalogue record for this book is available from
the National Library of New Zealand.
ISBN 978-1-877385-71-1

Design by Sorelle Cansino
Digital imaging by Jeremy Glyde
Printed by Everbest Printing Co. China

Front cover image: humpback whale, photo by Ingrid Visser,
Orca Research Centre
Front flap: sperm whales, illustration by Geoff Cox
Back cover: bottlenose dolphins in Fiordland,
photo by Barbara Todd

This book is dedicated to
J1 (1951–2010)

CONTENTS

PART ONE
ALL ABOUT WHALES

PART TWO
OF WHALES AND MAN

PART THREE WHALES AND DOLPHINS
OF AOTEAROA NEW ZEALAND

PREFACE

Nothing was visible as I dove into the sea, into a blue that deepened as it disappeared beneath me into an empty void. I soon realised, however, that I was not alone in that indigo space, for I could hear the songs of humpback whales reverberating around and through me. For a time, I hung there, rocked gently by the sea and the sounds of the singing whales. And then, out of the void, a dark shape appeared and started to drift towards me. As it came closer, the shape slowly metamorphosed and took on the features of a whale. When it finally surfaced, I was eyeball to eyeball with a humpback whose 4-metre flipper was just centimetres from my facemask, and tucked beneath that lengthy appendage was a tiny replica – a very young calf.

For a while, the calf and I stared into each other's eyes and then, growing bold, it left the safety of its mother's side and swam around me. Suddenly, although I saw no signal from the mother, the calf darted away as if summoned and returned to the shelter of the giant flipper.

For a few moments longer, the 15-metre mother calmly surveyed the tiny human that had entered her world.

I realised that should she swim off I could receive a mighty whack from her large tail. As if reading my mind, the mother and her calf slowly started to sink back into the deep blue from which they had appeared. As the pair moved away, a reverse metamorphosis took place and once again they became nothing more than a dark shape in the indigo void.

My encounter with the mother and her calf was poignant as the humpback is just one of the large whale species that humans have taken to the very edge of extinction. By the 1960s whaling had reduced global humpback populations to pitifully small numbers. As the mother calmly accepted me into her world, I thought of her ancestors… It's very possible that her mother was killed by my kind.

Those brief but unforgettable moments with the mother and calf started me on a journey that led to Aotearoa, and for over 25 years, I have had the privilege to encounter and work with many of the whale and dolphin species that inhabit the waters of Aotearoa and the South Pacific. This book tells a small part of their story.

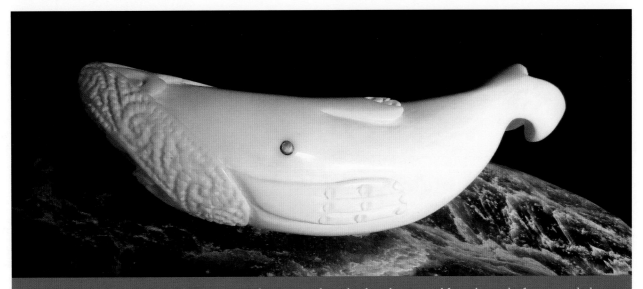

As I wrote this book, a small whale accompanied me – a tiny humpback replica carved from the tooth of a sperm whale. My humpback, or paikea as the species is known to Māori, swims on a sea of tumbled river pounamu (jade) and has a dorsal fin carved in the shape of a heru, a traditional Māori hair ornament. The whale's creator, Brian Flintoff, gave the carving to me to guide me on my journey writing about Aotearoa's whales and dolphins. I named my paikea Heru.

ACKNOWLEDGEMENTS

There are many people who have made this book possible, including hundreds of researchers from New Zealand and around the world who have dedicated their lives to discovering new information about cetaceans. It is impossible to name them all, but without their work, books such as this could not be written. I must, however, make a special mention of Michael Bigg, Ken Balcomb, Bernd Würsig and Alan Baker – mentors and friends who have inspired, encouraged and helped me along the way.

As well, I extend a huge thank you to Anton van Helden, Rochelle Constantine, Carlos Olavarria, Pamela Lovis and Ewan Fordyce, who, despite being incredibly busy, always had time to answer my queries and provide information and images. Thank you to everyone who worked on the exhibition *Whales | Tohorā* at the Museum of New Zealand Te Papa Tongarewa. I wish to acknowledge scientists Scott Baker, Nan Hausar, Ellen Garland, Michael Noad, Emma Carroll, Karen Stockin, Will Rayment, Phil Clapham, Yulia Ivashchenko and Nadine Bott who shared images, scientific papers and updated information on the species they are currently studying.

The artwork, illustrations and images have made this book a visual treat. Thank you to artists Brian Flintoff, Robin Slow, Metua Tangiiatua and all the others whose work features in the publication; to Geoff Cox for his amazing illustrations; and to Kim Westerskov and the many other individuals and organisations who shared their beautiful photographs.

Department of Conservation scientist Nadine Bott and the former whalers taking part in the annual Cook Strait Humpback Whale Survey are wonderful, as is Heather Heberley, who cheerfully takes care of everyone involved – thank you for making me feel so welcome and for your help.

Thank you to Hone Taumaunu of Ngāti Konohi, Te Warena Taua of Te Kawerau-a-Maki, Kukupa Tirikatene of Ngāi Tahu, Ramari Stewart of Ngāti Awa, former whalers Joe Heberley, Peter, Adrian, Ron and Ted Perano, Tom and Johnny Norton and Basil Jones, and others who have shared stories. I want to especially acknowledge Wade and Jan Doak, who have collected thousands of stories of human–cetacean encounters, stories which have given us another perspective of whale and dolphin consciousness.

Gratitude to organisations such as the Department of Conservation, Greenpeace New Zealand, WWF New Zealand and Project Jonah that work tirelessly to help protect and save whales and to individuals Mike Donoghue, Sue Miller Taei and the many others who have spent a large part of their lives working for the conservation of whale species. New Zealand universities such as Auckland, Massey and Otago have amazing programmes for cetacean studies, first class instructors and passionate students, many of whom go on to develop their own research programmes.

A big thank you to young conservationist Isaac Scott and to 'Save the Whales' artists James Sutherland and Meika Surgenor – education is the key to the survival of whales and their ocean homes.

Editor Susi Bailey is amazing – she made sense of it all and almost kept me sane. Thank you to Claire Murdoch, Sue Beaton, Harriet Elworthy, Angelique Tran Van and the Te Papa Press staff for all their hard work, and to everyone who worked on the book: Sue Hallas, Mike Donoghue, Rochelle Constantine, Ashley Remer, Sorelle Cansino, Catherine Cradwick, Louise Belcher, Claire Gummer, Martin Lewis, Hannah Newport-Watson, Daphne Lawless, Michael Hall and Jeremy Glyde.

And last but certainly not least, I extend a huge thank you to Roger, my family in New Zealand and in the USA, and friends who have put up with my limited attention while I researched and wrote this book.

INTRODUCTION

Whales and dolphins have captured the imagination of humans for centuries; they have been the subject of stories, fables, songs and poems; revered by some, feared by others and little understood by most. There are no fewer than 87 cetacean species on earth, one of which is the largest creature ever to have existed.

This book is a journey into the world of cetaceans – in particular, the world of the 42–45 species and subspecies that are known in New Zealand waters. It plunges into their ocean home and goes back millions of years to a time when the ancestors of whales were four-legged, hoofed creatures living on the edge of a warm tropical sea. This view into the whale's past entails, among other things, tracing the evolution of cetaceans over the millennia into creatures with physical attributes that sometimes exceed human capabilities and with social attributes that in many ways are like our own. Subsequent chapters examine some of the complex and ever-changing relationships that humans have had with cetaceans, which were sometimes treated as friends and other times as foes. Whales and dolphins have been seen as mystical beings that aid humans; as sea monsters to be feared; as a source of food, bone and oil that had both cultural and commercial value; as a symbol of environmental movements; and as a subject of scientific study.

The final part of the book looks more closely at each of the whale, dolphin and porpoise species that have been recorded in New Zealand waters. Some of these species are well documented, while others are still mysteries known from a few bone fragments or a skeleton discovered on a beach.

The link between man and nature has become increasingly fragile. It is my wish that this journey into the world of cetaceans will give knowledge about and evoke a sense of awe and appreciation for our fellow mammals. It must be remembered, however, that whole books have been written about each of the subjects covered, indeed about each of the species found within these pages. This introduction to New Zealand's whales and dolphins will hopefully inspire readers to discover more about these fascinating creatures, which like us are caught 'in the net of life and time, fellow prisoners of the splendour and travail of the earth'.

We need another, a wiser and perhaps a more mystical concept of animals. Remote from universal nature and living by complicated artifice, man in civilization surveys the creatures through the glass of his knowledge and sees thereby a feather magnified and the whole image in distortion. We patronize them for their incompleteness, for their tragic fate of having taken form so far beneath ourselves. And therein we err and greatly err, for the animal shall not be measured by man. In a world older and more complete than ours, they move finished and complete, gifted with extensions of the senses we have lost or never attained, living by voices we shall never hear. They are not brethren, they are not underlings; they are other nations caught with ourselves in the net of life and time, fellow prisoners of the splendour and travail of the earth.

Henry Beston, *The Outermost House* (1928)

PART ONE

ALL ABOUT WHALES

*They say the sea is cold, but the sea contains
the hottest blood of all, and the wildest, the most urgent.*

...

*The right whales, the sperm-whales, the hammer-heads, the killers
there they blow, there they blow, hot wild white breath out of the sea!*

D.H. Lawrence, 'Whales Weep Not' (1932)

THE
SALTY SEA

If there is magic on the planet,
it is contained in water.

Loren Eiseley, *The Immense Journey* (1957)

We share our planet with no fewer than 87 different whale, dolphin and porpoise
species. Although a few dolphin species have adapted to life in fresh water,
the majority of these air-breathing marine mammals make their home in salty
seas. A view of Earth from space reminds us that our planet is dominated by this liquid
environment – the world's seas and oceans cover more than 70 per cent of its surface, while
the largest and deepest, the Pacific, alone covers almost 30 per cent. This mighty ocean
plunges to a depth of almost 11 kilometres in the Mariana Trench, and to over 10 kilometres
in the Kermadec Trench between New Zealand and Tonga. Altogether, Earth's oceans
contain more than 1.3 billion cubic kilometres of water and host millions of living creatures,
ranging from microscopic organisms to the largest animal ever known to have lived,
the blue whale.

THE EVER-CHANGING SEAS

The outer layer of Earth is like a jigsaw, made up of several sections called tectonic
plates. These plates move around, sometimes colliding, sometimes moving apart
and sometimes sliding past one another. Over the millennia, the face of the
planet has changed as the plates and the land they carry have come together to form
'supercontinents' or drifted apart to form new continents. As the land masses shifted, they
affected the waters surrounding them; at times, new oceans were opened up, while at others,
existing oceans were closed off and disappeared.

*Earth was originally devoid of large bodies of water – it was once, in fact, a 'brown' planet.
The salty seas started to form around four billion years ago, but it took millions of years for
them to evolve into the oceans we know today. The present-day ocean floor is pictured here.*

CAMBRIAN PERIOD
(500 MILLION YEARS AGO)

At this time the Panthalassic Ocean surrounded the supercontinent Pangaea and covered most of the northern hemisphere. Another ocean, Iapetus, separated Laurentia (North America) from Baltica (Europe), while the supercontinent Gondwana lay to the south. The seas abounded with life, with simple animals evolving to become more complex and diverse.

CARBONIFEROUS AND TRIASSIC PERIOD
(350–200 MILLION YEARS AGO)

Continental land masses stretched from pole to pole, almost encircling the Palaeo-Tethys Sea, and the Panthalassic Ocean now lay to the west of Pangaea. Large portions of Gondwana extended over the South Pole and were covered by the southern ice cap. Fish were abundant but life on land had increased too, with amphibians initially dominant and reptiles developing from them.

JURASSIC PERIOD
(200–150 MILLION YEARS AGO)

Parts of what would become Asia broke away from Gondwana and moved north, opening up the Tethys Ocean. The central Atlantic Ocean started to form, splitting Pangaea into northern and southern continents. The worldwide climate heated up as warm equatorial currents spread north and south, resulting in the loss of the southern ice cap. This was the age of the dinosaurs on land, and also saw the appearance of the first mammals.

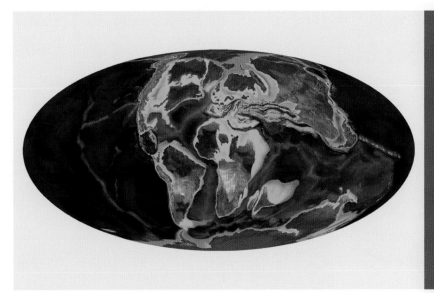

CRETACEOUS PERIOD
(150-65 MILLION YEARS AGO)

Gondwana continued to break apart and India and Africa started to drift away. The Tethys Ocean became smaller as the Atlantic Ocean continued to expand. Europe separated further from North America, while the Arctic Ocean opened up over the North Pole. On land, dinosaurs continued to dominate, although this time also saw further diversification of mammals.

PALAEOGENE PERIOD
(65-25 MILLION YEARS AGO)

India continued to drift north, ultimately colliding with Asia, resulting in the uplift of the Himalayas. The Atlantic Ocean became firmly established, while the Tethys Ocean shrank. Around 55 million years ago, the land ancestors of today's whales and dolphins began their first forays into the Tethys. By the time the ocean had disappeared completely, these creatures were fully aquatic and had moved into other large bodies of water.

NEOGENE PERIOD
(25-2.5 MILLION YEARS AGO)

When our oceans arrived at their present form, around 15 million years ago, they contained the majority of the whale, dolphin and porpoise species we know today. The oceans will continue to change through tectonic processes as they have throughout Earth's history,

COOL CHANGE

Around 65–40 million years ago, Earth was much warmer than it is today. There were no polar ice caps as warm equatorial ocean currents moved freely, carrying heat to higher latitudes. Fully aquatic whale and dolphin species were starting to spread into other bodies of water in search of food as the Tethys Ocean decreased in size. Around 35 million years ago, however, Gondwana's final break-up created dramatic climate changes that affected the ecology of the southern oceans. When the Australian continent broke away and moved northward, a cool circumpolar Antarctic current was created, blocking the warm waters flowing from the Equator. Near-freezing water temperatures developed around Antarctica and the polar ice cap began to re-form. As the cool circumpolar current travelled eastward, it converged with the warm westward-moving northerly currents, creating nutrient-rich upwellings that supported diverse new communities of oceanic life. During this period of change, some marine species became extinct but other species, including the ancestors of our modern whales, evolved and flourished as abundant food resources became available.

THROUGH THE OCEAN DEPTHS

Since life first appeared in our oceans around 3.5 billion years ago, countless species have evolved to fill every marine habitat, from the surface waters to the deepest trenches. Life varies greatly beneath the surface of the sea. The deeper you go, the darker it gets and the more bizarre the life forms become. Only a small proportion of marine species inhabit the deepest zones, with most living above 1000 metres and the majority occurring in the upper 200 metres, where light still penetrates. Over millions of years, whales and dolphins have adapted to feed at different depths. Dolphins and baleen whale species feed closer to the surface, while the deepest diver, the sperm whale, sometimes feeds at depths that exceed 2000 metres.

Like outer space, the vastness of our oceanic 'inner space' can be overwhelming. In order to simplify and categorise life in the sea, scientists have divided the ocean into a series of distinct zones, each of which provides different living conditions and hosts different species that have adapted to meet those conditions.

The deepest diving cetacean, the sperm whale, has been tracked to depths of 2800 metres.

EPIPELAGIC ZONE (0–200 METRES)

The basis for all life in our seas is plankton, which are (usually) microscopic organisms that sit at the base of the oceanic food web. While some plankton (from the Greek, meaning 'that which drifts') are weak swimmers, most forms do not swim at all and are merely carried by the ocean's currents. All forms of phytoplankton (plant plankton) adhere to a 'live fast, die young' philosophy, making the most of a very short life by either doubling or quadrupling their numbers daily. Photosynthesising phytoplankton occur in massive swarms in the epipelagic or sunlit zone, the top 200 metres of the ocean, during spring and early summer. The phytoplankton blooms attract vast hordes of zooplankton (animal plankton), along with many other species that feed on them.

The most visible animals in the sunlit zone are fish species and other marine vertebrates such as seals, whales and dolphins, although marine invertebrates such as corals, sponges, worms, starfish, jellyfish, octopus, squid and molluscs all play an equally important role. Baleen whales, along with most dolphin species, feed mainly in the epipelagic zone.

A RICH ENVIRONMENT

The top metre of ocean water is the most nutrient-rich layer and also the place where gas exchange occurs between the ocean and the atmosphere. Up to 50 per cent of the oxygen in Earth's atmosphere is produced by phytoplankton, which utilise sunlight to photosynthesise, capturing solar energy and transforming it into chemical energy and releasing oxygen in the process. The upper layer of the sea is vitally important to all life on the planet but is also the most susceptible to many forms of pollution.

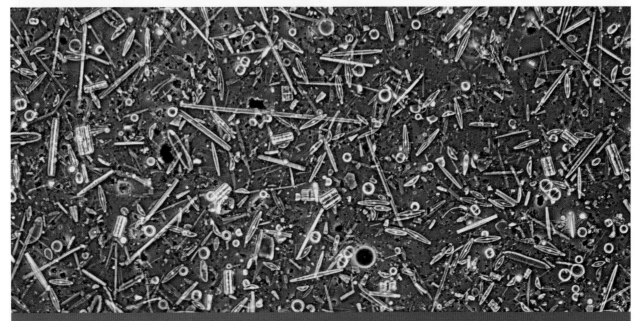

This diatom mixture represents one of the most common forms of phytoplankton. There are an estimated 100,000 extant diatom species, most of which are single-celled algae that are encased within a cell wall composed of silica.

ZOOPLANKTON

Zooplankton, or animal plankton, comprise hundreds of species, including worms, sea squirts and salps, and crustaceans such as amphipods, copepods and krill. In addition, the larvae of many coral, shellfish, crab, starfish and fish species are zooplankton temporarily before metamorphosing into their adult forms. For a whale, copepods and krill are two of the most important forms of zooplankton in the sea.

Copepods

Copepods are tiny organisms with a minuscule body about the size of a grain of rice. When they are feeding, they hold their long antennae out to the side, giving them a T-shaped appearance; the antennae are also used as an anchor to enable these quick-moving crustaceans to stop suddenly.

Although they are small, copepods are extremely numerous – 1 cubic metre of sea water may contain more than 20,000 individuals.

Most copepods are herbivores that feed on phytoplankton, but some feed on zooplankton, including other copepod species. Almost all species are bioluminescent, and at night the trails of dolphins are often seen as a sparkling display of flashing light created by the bodies of thousands of copepods in the water streaming off the dolphins' bodies. Copepods are one of the favoured foods of skimming right whales, as well as being a prey source for other baleen whales and many other marine species. Although they are consumed by the millions, they breed so rapidly that they are able to keep up with the gastronomic demands of their predators.

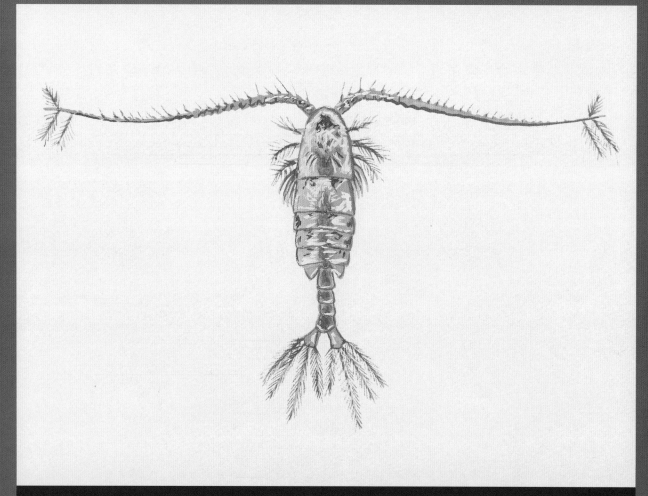

The 4-millimetre-long Neocalanus tonsus *is a relatively large copepod. In the southern hemisphere, the species' vast range stretches from subantarctic waters all the way to the subtropics.*

Krill

Krill resemble shrimp but are smaller and belong to a separate genera. All but one of the world's 87 known species of the krill family Euphausiidae is bioluminescent, with rows of photophores along their bodies that emit blue flashes of light. One New Zealand species, the transparent *Nyctiphanes australis*, often washes ashore during summer months, producing spectacular light shows on the beaches where they have been stranded. One of the largest and best-known krill species, *Euphausia superba* or Antarctic krill, is the main food source of the Southern Ocean's whale, seal and seabird populations. Summertime swarms of this krill have been known to reach densities of an estimated 30,000 individuals or more per cubic litre of sea water. Antarctic krill can moult or shed their skin in seconds and will sometimes do so en masse. When they are frightened, swarms of krill will literally jump out of their skins.

Euphausia superba, or Antarctic krill, are up to 60 millimetres long and live up to six years.

Nyctiphanes australis, *or transparent krill.*

FASCINATING FACT

In terms of biomass, *Euphausia superba* (Antarctic krill) is possibly the most abundant animal species on the planet.

MESOPELAGIC ZONE (200–1000 METRES)

Also known as the twilight zone, this is a region of gloom where light has all but disappeared and creatures with large eyes play a constant game of hide and seek. In order to avoid detection, most twilight animals are slim and silvery or almost transparent, and over 90 per cent are bioluminescent, producing light for communication, courtship, camouflage and to attract prey. The mesopelagic zone contains numerous mid-water squid species, which are one of the main food sources for sperm, pilot and beaked whales. Some species hide in the twilight zone during the day and then move upward into the epipelagic zone at night to feed.

ONE OF EARTH'S GREATEST MIGRATIONS

Every night, millions of marine animals make a vertical migration upwards in order to feed in the nutrient-rich waters of the epipelagic zone. Small zooplankton may travel only 10-20 metres up to feed on surface phytoplankton, while larger animals may rise more than 1000 metres. As the sun comes up, most descend back to the relative safety of the gloomy twilight of the mesopelagic zone. Two of the most abundant fish species involved in this vertical migration are lanternfish, named for the light-producing photophores that cover their bodies, and bristlemouths. These fish journey from depths as great as 1700 metres up to 100 metres, and then make the return trip at dawn.

New Zealand's dusky dolphins feed mainly at night and take advantage of this upwards migration – one of their favourite meals is lanternfish.

Many people picture sperm whales battling with giant squid for every meal, but in truth the majority of their feeding takes place in the mesopelagic zone and consists of much smaller (2–4-metre) squid species (top, right) along with ling (bottom), grouper (top, left) and other mid-water fish species.

BATHYPELAGIC ZONE (1000–4000 METRES)

Within this so-called midnight zone, darkness is broken only by fleeting flashes of bioluminescent light emitted from its deep-sea inhabitants. Temperatures are chilly, averaging just 1–2°C, and only highly adapted animals can survive the extreme pressure – some have liquid-filled bodies that are almost incompressible. Creatures must conserve energy in this cold, oxygen- and nutrient-poor environment, and as a result most species are large, sluggish and sedentary – underwater couch potatoes! Food is scarce, so meals are rare and unpredictable. Species often employ a 'bait and catch' technique, whereby they flash bioluminescent light and then catch the curious prey attracted to it.

This is the home of the giant squid, one of the prey species of the deep-diving sperm whale.

Many bathypelagic species, such as anglerfish (left) and gulper eels (right), have huge heads, terrifying sets of teeth and expandable stomachs that allow them to catch prey at least as big as themselves.

DEEP-SEA WHALE FOOD

Very few whale species can dive to depths greater than 1000 metres, but sperm whales – along with a few beaked whale species – are the exception. Sperm whales are the deepest divers and have been tracked to depths greater than 2000 metres. The main attraction is deep-sea squid, some species of which are rather enormous!

The giant squid *Architeuthis dux* (right) is one of the world's largest squid species and one of the largest invertebrates – an individual caught in a deep-water trawl measured over 16 metres in length and weighed more than 800 kilograms. When scientists examined the stomach contents of a sperm whale that stranded in March 1996 on a Paekakariki Beach in New Zealand's North Island, they discovered more than 3000 squid beaks from 24 different species, including giant squid.

Sperm whales also feed on deep-sea sharks – in fact, the remains of *Scymnodon* species, bottom-dwelling sharks that live at about 3000 metres, have been found in the stomachs of sperm whales. Whether the whales are diving to that depth or the sharks are coming up into less deep waters is not yet known.

ABYSSOPELAGIC ZONE
(4000 METRES TO THE OCEAN FLOOR)

Also known as the abyssal or lower midnight zone, these inhospitable waters extend down to the seabed at around 4000 metres. Half of our ocean floors are covered by abyssal plains: flat, featureless expanses that together form one of the largest single environments on Earth. Only a tiny fraction of this area has been seen or explored.

Much of the abyssal plains are covered with fine sediments and appear almost lifeless until you look closer and discover that there is, in fact, life everywhere, camouflaged or hidden within mounds and small trenches. Sea cucumbers are one of the most dominant life forms but they are joined by a surprising variety of other animals, including brittle stars, sea urchins, worms, sponges, sea anemones, bivalve molluscs and foraminiferans (single-celled organisms that build intricate homes out of mud and pieces of shell).

Many of the animals that inhabit this zone feed on 'marine snow', minuscule pieces of organic matter that have drifted down from the upper layers of the sea. Others have adapted to life around oceanic hydrothermal vents, where chemical-rich geothermally heated water is emitted from chimney-like structures at geologically active sites. There, bacteria use the chemicals in the water to create organic matter, which in turn supports a whole host of other organisms.

FASCINATING FACT

The death of a large creature such as a whale provides a banquet for deep-sea dwellers — scientists have found an estimated 12,000 animals from 43 species feeding on the remains of a single whale.

One of the most abundant abyssopelagic fish is the rat-tail, which resemble overgrown tadpoles (bottom). The most ferocious looking, the fangtooth fish, have large heads and massive teeth (top). Fangtooth fish have been recorded at depths of almost 5000 metres.

HADOPELAGIC ZONE (4000–11,000 METRES)

The abyssal plains are broken up by huge mountain chains or mid-ocean ridges, which cross all the major oceans except the North Pacific. In other areas, the seabed plunges into deep trenches, known as hadopelagic or hadal zones, where depths can reach 11 kilometres. Very little is known about life in the hadal zone, which is well beyond the reach of any whale species and almost beyond that of humans.

The waters of the hadal zones were once believed to be devoid of life, but deep-sea expeditions have discovered that in fact they are teeming with organisms, including fish, snails, starfish, shrimps, sea cucumbers, worms and molluscs, many of which have only recently been discovered. More than 100 species have been recorded from New Zealand's Kermadec Trench, one of the deepest in the world. The trench is home to the Kermadec snailfish, a gelatinous tadpole-shaped fish known only from five specimens that were trawled up from a depth of 6700 metres. The cusk eel, a relative of the common ling, lives at even greater depths, more than 8 kilometres below the surface. Like animals inhabiting the abyssopelagic zone, many of these species are necrophages, feeding on dead material – including cetaceans – that descends from higher levels.

FASCINATING FACT

In 1960, the deep sea vessel *Trieste* descended 10,911 metres into the Mariana Trench. Upon reaching the bottom, the two crew looked out of their porthole and found themselves staring at a small flatfish. Their record dive wasn't equalled until 42 years later, when film director James Cameron made a solo descent into the Mariana Trench on 26 March 2012. When Cameron reached the trench floor, his 9.6-metre sub had shrunk by almost seven centimetres due to the incredible force of pressure exerted on it.

In 2011 a 'supergiant' amphipod, Alicella gigantea, was captured in the Kermadec Trench. The 28-centimetre-long specimen was found at a depth of six kilometres by a team of researchers from Scotland, Japan and New Zealand. NIWA scientist Ashley Rowden is pictured with it here.

FROM TINY PLANKTON TO MIGHTY WHALES

Plankton are the foundation for all life in the sea. Phytoplankton are consumed by zooplankton, which in turn are a food source for hundreds of species ranging from fish, squid, sharks, penguins and other seabirds, to some seal species and baleen whales. The gastronomic pecking order also includes small fish and squid that are eaten by larger species, while fish, squid and octopus of various sizes are consumed by seals, whales and dolphins. This continuous 'all-you-can-eat buffet' is driven by ocean currents, which carry and mix nutrients throughout the sea.

The oceanic food web sounds reasonably simple when described like this but it is in fact one of the most intricate and tangled food webs in existence. Even top predators contribute by hosting other living creatures such as barnacles or amphipods, as well as through the decomposition of their own bodies when they themselves die.

Although whale barnacles are referred to as ectoparasites, in reality the crustaceans do not feed on their hosts, but instead use the whales as a moving platform from which to strain passing plankton.

AOTEAROA NEW ZEALAND'S SEASCAPE

The birth of Aotearoa began around 100–80 million years ago, when rifts appeared on the southeastern corner of Gondwana and a newly formed landmass, Zealandia, started to separate from the supercontinent. Over the next 30–40 million years, Zealandia was pushed and pulled up, down and sideways as its two main tectonic plates shifted and moved, finally resulting in its current form around 12–5 million years ago.

Today, the country has one of the largest marine environments in the world. Its islands are surrounded by more than 4.2 million square kilometres of water, or more than 14 times the size of its total land area. Its Exclusive Economic Zone (EEZ) is the fifth largest in the world,

stretching almost 3000 kilometres from the subtropical Kermadec Islands in the north at 30°S, to the subantarctic Campbell Island at 52.33°S.

Beneath the surface of New Zealand's seas is a fascinating and varied marine world. Among its more exotic features are enormous seamounts – submerged, usually volcanic mountains that rise thousands of metres above the sea floor. Interspersed among these dramatic features are large plateaux dotted with hills and valleys, deep submarine canyons and meandering river-like channels. Abyssal plains, occur at average depths of 4000 metres. These large, flat areas of the ocean floor extend across almost a quarter of New Zealand's seabed and are covered

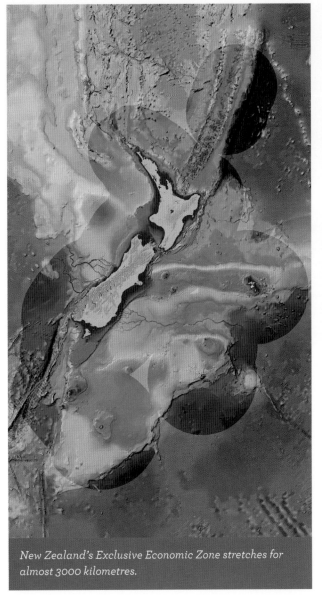

New Zealand's Exclusive Economic Zone stretches for almost 3000 kilometres.

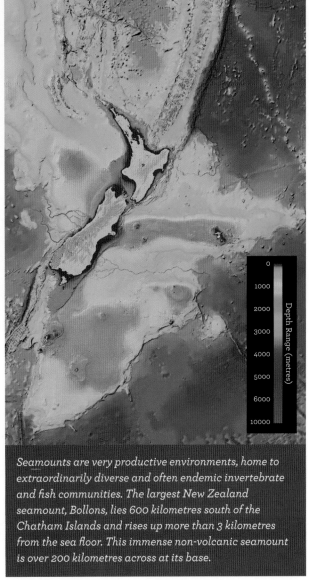

Depth Range (metres)

0
1000
2000
3000
4000
5000
6000
10000

Seamounts are very productive environments, home to extraordinarily diverse and often endemic invertebrate and fish communities. The largest New Zealand seamount, Bollons, lies 600 kilometres south of the Chatham Islands and rises up more than 3 kilometres from the sea floor. This immense non-volcanic seamount is over 200 kilometres across at its base.

This migrating humpback whale was observed off New Zealand's coast during its annual sojourn to breeding grounds in the South Pacific.

with a thick blanket of sediments. And plunging to mind-boggling depths of 10 kilometres is the 1500-kilometre-long Kermadec Trench.

Inhabiting this seascape is an equally diverse marine life. At least 42 of the world's population of whale, dolphin and porpoise species, along with at least three subspecies, have been observed in New Zealand's waters; some are resident, while others either migrate through or visit temporarily. However, the sight of a pod of playful dolphins or a leaping humpback whale sometimes makes it easy to forget that the number of cetacean species is actually minuscule compared to the many thousands of other marine species in the ocean environment.

Scientists estimate New Zealand's seas may support more than 17,000 marine species, if numbers of known inhabitants are combined with the thousands of unnamed specimens that have been collected but are yet to be described. Each plays a significant role in the ocean's complex food web – like the web of a spider, each small thread is essential, and if just one is damaged or destroyed the web is weakened and its efficiency altered. Every individual species in our seas, from the tiniest picoplankton (smaller than the width of a hair), to the largest blue whale, is part of this food web and is vital to the overall health of the ocean. Some species live in harmony, others devour each other, but everything is connected.

New Zealand's long isolation from other marine ecosystems has resulted in a high rate of endemism. At least 44 per cent of its known oceanic species, including one cetacean, the Hector's dolphin, do not live or breed anywhere else in the world.

IN THE BEGINNING

*Diviner than the dolphin is nothing yet created;
for indeed they were aforetime men and lived in cities along
with mortals, but by the devising of Dionysus they exchanged
the land for sea and put on the form of fishes.*

Oppian of Corycus, *Halieutica* (third century AD)

Whales, dolphins and porpoises are mammals, but they are so streamlined and well adapted for life in the sea that for centuries they were referred to as fish. Even the Greek philosopher and scientist Aristotle, who recognised their mammalian characteristics, referred to dolphins as 'air-breathing fishes in the sea'. Their status as a mammal finally became official in 1758 when Carolus Linnaeus, the Swedish naturalist, solemnly declared, 'I hereby separate the whales from the fish.' Whales, dolphins and porpoises were placed in the order Cetacea (from the Greek ketos, meaning 'sea monster') and became collectively known as cetaceans. For years, one of nature's great unsolved mysteries has been the origin of whales. What were their ancestors? Where did they come from? And how did they evolve?

About 360 million years ago, amphibious creatures in the form of salamander-shaped tetrapods came close to land. These semi-aquatic tetrapods were able to breathe air and had limited locomotion which enabled them to move in very shallow waters. Over the millennia, these animals developed the ability to walk on land. For some unknown reason, around 55 million years ago, a few of these terrestrial mammals did an evolutionary about-face and returned to the sea to become the whales we know today.

The early inshore tetrapods took advantage of a newly evolved and rich semi-aquatic ecosystem created by the emergence of vegetation. These wetlands provided an abundant food source, which included plant life and insects.

FOSSIL FIRSTS

F or years, large gaps in the fossil records of whales and the history of their evolution remained a mystery. In the 1960s, Leigh Van Valen, a palaeontologist at the American Museum of Natural History, found a 'striking resemblance' between the teeth from the early known fossil whales (the archaeocetes) and those from a family of extinct meat-eating creatures known as Mesonychidae. This discovery led scientists to believe that whales may have descended from carnivorous wolf-like animals such as the mesonychids, which lived on the edge of warm swamps and other waterways around 55-50 million years ago. To prove the theory, scientists now needed fossil remains linking this specific group of land mammals with whales.

The first big breakthrough occurred in the mid-1970s, when vertebrate palaeontologist Philip Gingerich began working in a desert in northwest Pakistan that contained 50-million-year-old fossils from the ancient coastal beds of the Tethys Ocean. The team was not looking specifically for whale fossils, but they did start to uncover some strange whale clues. In 1977, they found pelvic fragments that they joked must have belonged to a 'walking whale'. Then in 1979, the team uncovered a partial wolf-sized skull that had distinctive cetacean characteristics; they called the animal *Pakicetus*. These discoveries led to confirmation of proposals made in the 1800s that an ancestor of modern-day whales had indeed once walked on land.

Palaeontologist Philip Gingerich and his team named their 50-million-year-old whale ancestor Pakicetus, *a combination of Paki- (from Pakistan) and -cetus (whale). This carnivorous mammal is thought to have hunted for food in rivers flowing into the Tethys Ocean.*

The tiny legs of the long, snake-like Basilosaurus *would have been unable to support the whale on land, but they still had attachments for powerful muscles, complex locking mechanisms in the knee and functional ankle joints.*

In the 1980s, Gingerich shifted his investigation area from Pakistan to a part of Egypt known as Wadi Al-Hitan. This desert contained formations around 10 million years younger than those in which *Pakicetus* was discovered. The team excavated numerous partial skeletons of fully aquatic 37–40-million-year-old whales such as the 16-metre *Basilosaurus* and the 5-metre *Dorudon*. These ancient whales had hollow ear bones, along with other adaptations for underwater hearing, and streamlined bodies with elongated spinal columns and powerful tail flukes.

In 1989, the team made their most astounding discovery of all when they uncovered a whale knee at Wadi Al-Hitan. Other pieces of hind limbs followed: a femur, a tibia and fibula, then an ankle and foot. The final culmination came with the discovery of a complete set of long, slender toes. It must have seemed surreal to sit in a dry, sandy desert reconstructing the hind legs of a whale species that swam there over 40 million years ago. The hind limbs of the 16-metre *Basilosaurus* they had found would have been about the size of the legs of a three-year-old child.

Wadi Al-Hitan Valley, located in the western Egyptian Desert, contains a high concentration of fossils from some of the youngest archaeocetes (ancient whales). These fossils clearly demonstrate early whales in the last stages of losing their hind limbs during their final transition from land to sea.

Ambulocetus natans, *the most primitive known saltwater cetacean, appears to have lived between land and sea. The species is known only from skeletal remains found in near-shore marine environments such as bays and estuaries.*

The discovery was a huge breakthrough, linking whales to terrestrial ancestors that had walked on land. But there were still some 10-million-year gaps in the fossil records between *Pakicetus* and the fully aquatic *Basilosaurus*.

A huge part of these gaps was filled in 1992, when Hans Thewissen discovered the near-complete skeleton of a 45–48-million-year-old whale, which he named *Ambulocetus natans*, meaning 'walking, swimming whale'. *Ambulocetus* was adapted for both terrestrial and aquatic environments, with sturdy leg bones and flexible elbow and wrist joints for walking on the land, and large paddle-like feet and a powerful tail for swimming in the sea.

A well-preserved whale jaw (left) protrudes from a cliff in Wadi Al-Hitan.

THE CRADLE OF WHALES

Around 60–55 million years ago, the Tethys Ocean stretched across a vast region between modern-day Gibraltar and India (right). The warm, shallow waters teemed with life, including numerous fish species as well as ancient sharks and crocodiles, and it was here that the early whale ancestors made their return to the sea. The eastern Tethys closed off around 34 million years ago, eventually leaving in its place a vast desert region in what is now India and Pakistan. At first glance, the remnants of this magnificent landscape appear quite barren, but a closer look reveals a treasure trove of ancient marine fossils buried within the rocks and sand. Each buried fossil tells a story and reveals answers that help to solve the mystery of its liquid past.

KEY FINDS

Since Thewissen's 1992 discovery, he, Gingerich and others have unearthed many more fossils documenting additional stages of the transition of whale ancestors from land to sea. The fossils demonstrate that as these archaic whales evolved, they became less terrestrial and more aquatic; their bodies became sleek and streamlined as forelimbs became flippers and powerful tail flukes appeared. Gradually, the nostrils migrated to the top of the head to become blowholes, while the eyes moved to the side to allow lateral vision. Internal changes also occurred, particularly within the inner ear.

For years, Leigh Van Valen's hypothesis that the meat-eating, wolf-like mesonychids were the terrestrial ancestors of whales was the generally accepted theory of whale evolution. However, this was questioned in the 1990s when DNA studies indicated that whales were linked more closely to the vegetarian Artiodactyla – even-toed ungulates such as camels, cattle, pigs and hippos.

Initially, palaeontologists like Gingerich and Thewissen were uncertain about the new theories, but ultimately they themselves found physical evidence supporting the artiodactyl hypothesis. The ankle bone, or astragalus, of all even-toed ungulates is a distinctive feature, with an unusual double-pulley shape that allows for greater spring and flexibility. In the early 2000s, palaeontologists working for both Gingerich and Thewissen found whale ankle bones with this shape, adding weight to the theory that whales are, after all, more closely related to grazing artiodactyls than to wolf-like carnivores.

A vast amount of knowledge has been gained about the evolution of whales through fossil discoveries and modern scientific techniques, but questions remain. For example, do the wolf-like Mesonychidae still fit into the evolutionary puzzle and were they somehow related to artiodactyls in the ancient past? And what was the common ancestor of the whales and much later hippos? In addition, there continue to be missing links between the ancient archaeocetes and many of our current cetacean species. The sands of time have provided many answers, but there are still pieces of the puzzle waiting to be unearthed. The final chapters on the whales of the past have yet to be written.

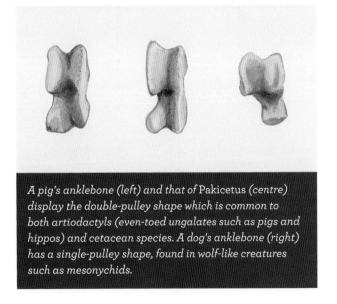

A pig's anklebone (left) and that of Pakicetus *(centre) display the double-pulley shape which is common to both artiodactyls (even-toed ungalates such as pigs and hippos) and cetacean species. A dog's anklebone (right) has a single-pulley shape, found in wolf-like creatures such as mesonychids.*

THE ANCIENT WHALES

All ancient whales are grouped in the sub-order Archaeoceti. Within this, scientists have discovered five different families containing at least 44 species. Most of these ancient whales had hind limbs that allowed them to function on land, but by the late Eocene period (56 to 34 million years ago)

the basilosaurids had become fully aquatic. The majority of the ancient whales became extinct towards the end of the Eocene period when dramatic climatic changes occurred, but evidence from New Zealand indicates that a few basilosaurids may have survived into the Oligocene period (33 to 23 million years ago).

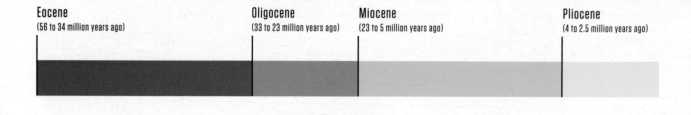

Eocene
(56 to 34 million years ago)

Oligocene
(33 to 23 million years ago)

Miocene
(23 to 5 million years ago)

Pliocene
(4 to 2.5 million years ago)

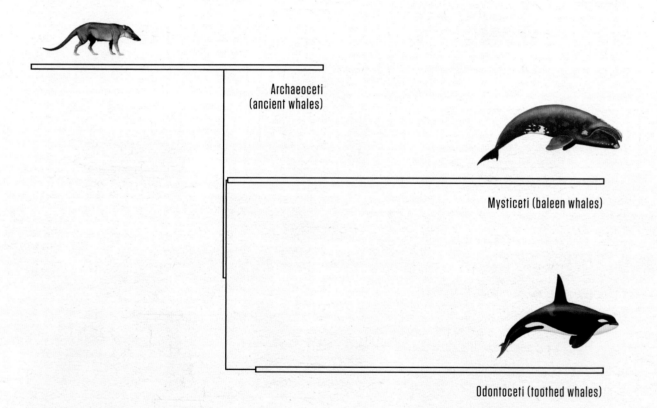

Archaeoceti
(ancient whales)

Mysticeti (baleen whales)

Odontoceti (toothed whales)

THE FIRST WHALE FAMILY PAKICETIDAE: 3 GENERA, 4 SPECIES

PAKICETUS ATTOCKI: 1.8 metres long, lived 55–50 million years ago

Pakicetus is known from dozens of fossils discovered in Pakistan, although a complete skeleton has never been found. Although it was primarily a terrestrial mammal, its tooth and jaw structure is consistent with a fish-eating diet. *Pakicetus* fossil remains have been found only in freshwater deposits believed to have been shallow streams. Their link to cetaceans is seen in adaptations of ear bones (the auditory bullae) that are unique to whales and dolphins.

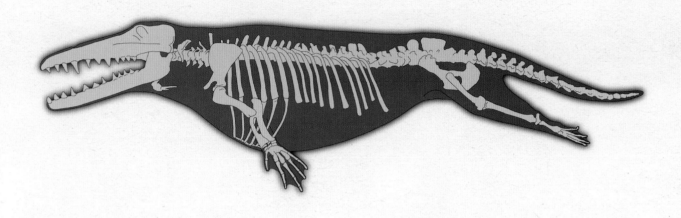

AMBULOCETUS NATANS: 4.2 metres long, lived 50–48 million years ago

Ambulocetus natans (above) is known from fossil discoveries in northern India and Pakistan, including a near-complete skeleton. With a scientific name meaning 'walking, swimming whale', this 4.2-metre-long animal had large feet and a powerful tail, and lived both on land and in the sea, although it probably came ashore to breed. *Ambulocetus natans* lived in shallow, swampy environments, and is likely to have hunted its food in an ambush fashion similar to modern-day crocodiles.

FAMILY REMINGTONOCETIDAE: 5 GENERA, 8 SPECIES

KUTCHICETUS MINIMUS: 1.75 metres long, lived 46–43 million years ago

This early whale had relatively long, slender jaws and an otter-like body with small hind legs and a long tail, and spent more time in the sea than did the ambulocetids. The hair that was a feature of its more terrestrial ancestors had all but disappeared; instead, blubber provided insulation and aided swimming. Remingtonocetidae fossils are known only from India and Pakistan; one of the oldest fossil members of the family, *Attockicetus*, was found in a site alongside *Ambulocetus*, indicating that some remingtonocetes also lived in near-shore environments such as bays and saltwater swamps.

MARINE EXPLORERS FAMILY PROTOCETIDAE: 11 GENERA, 14 SPECIES

RODHOCETUS SPECIES: 3 metres long, lived 44–40 million years ago

Belonging to the Protocetidae family, *Rodhocetus* species had compressed lower arm bones and long, delicate feet that were probably webbed. Fossils have been found in Indo-Pakistan, Africa, Europe and North America, and indicate that Protocetidae lived in shallow marine environments. Although the protocetids were the first to disperse across the oceans, it is believed that they reached those areas by travelling in shallow coastal waters rather than through the open ocean.

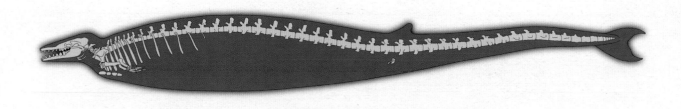

BASILOSAURUS CETOIDES: 16–18 metres long, lived 41–35 million years ago

The Basilosauridae are believed to be the progenitors of modern odontocete and mysticete whales. *Basilosaurus cetoides* (above) was a fully aquatic cetacean whose forelimbs had become paddle-like flippers but that still retained small, rudimentary hind legs. It had a long, snake-like body, a pelvis that was detached from the spinal column, and a tail flattened into tail flukes (the basilosaurids were the first groups to display this feature). The nostrils had begun their migration towards the top of the head to become a blowhole.

FAMILY BASILOSAURIDAE, SUB-FAMILY DORUDONTINAE: 6 GENERA, 9 SPECIES

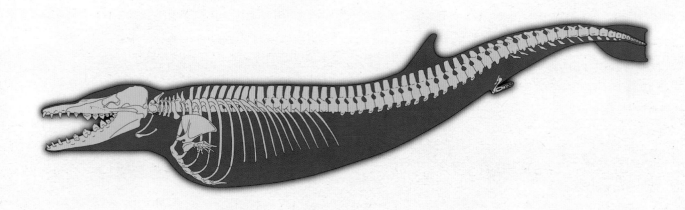

DORUDON ATROX: 4.5 metres long, lived 40–35 million years ago

The smaller basilosaurids, like *Dorudon atrox* (above), were more dolphin-like in their appearance. Although the majority of *Dorudon* fossils have been found in the eastern USA and Egypt, it is believed that these fully aquatic archaeocetes eventually went on to inhabit all the oceans of the world. Dorudon-like whales lived in New Zealand waters about 39 million years ago. These are some of the oldest reported members of the Basilosauridae family.

Modern forms of dolphins appeared
12–15 million years ago.

Sperm whales appeared around 25 million years ago.

THE RISE OF MODERN CETACEANS

The final transition from basilosaurids to modern whales was completed around 34 million years ago, when dramatic changes on Earth cooled its hothouse climate. The break-up of the supercontinent Gondwana resulted in the formation of a large ocean around Antarctica and the creation of a cool westerly circumpolar current. Lowered sea temperatures led to new areas of upwelling, bringing nutrients to the surface and attracting seasonal swarms of plankton, which in turn provided food for a profusion of sea creatures, including whales. In response to the new climatic conditions and abundant food sources, whales flourished and developed adaptations for new feeding techniques and strategies.

The two modern cetacean groups Odontoceti (toothed whales) and Mysticeti (baleen whales), had appeared by 34 million years ago, with evidence suggesting they may have evolved from *Dorudon*-like archaeocetes. Baleen (plates of bristle-like filters, used to trap food) proved to be an effective adaptation, enabling whales to feed on huge swarms of tiny prey. As a result, some archaic whales lost their teeth altogether and relied solely on baleen. Another group of toothed species developed echolocation, a biological sonar, to help locate prey, and some increased their ability to dive deeply in search of food.

The majority of our modern cetacean families were present in the world's oceans by the end of the Miocene period, around 5 million years ago.

WHALE FOSSILS IN AOTEAROA

Cenozoic-era rocks in Aotearoa dating back 39 million years have yielded a diverse range of fossil whale, dolphin and porpoise species. Some of the most significant finds have come from North Otago and South Canterbury in the South Island, and from the Wairarapa and the Wanganui Basin in the North Island. Although many of the specimens are fragmentary, there is enough material to determine that Archaeoceti (ancient whales) occupied the region before the rise of the now living Odontoceti (toothed) and Mysticeti (baleen) cetaceans, which together form the recently named group Neoceti.

Archaeoceti fossils found to date in Aotearoa are later forms belonging to the Basilosauridae family, and included the large Basilosaurinae, along with the smaller, more dolphin-like dorudonts. Palaeontologist Ewan Fordyce of the University of Otago has collected a number of fossils that have filled many gaps in the on-going story of whale evolution.

The earliest whales with baleen only were the eomysticetids. *Mauicetus lophocephalus*, collected by Fordyce from New Zealand rock formations estimated to be around 26 million years old, belongs to the extinct family Eomysticetidae.

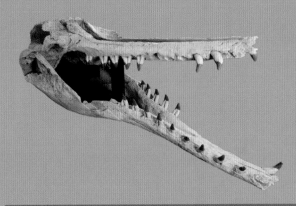

Squalodon *species lived around 25 million years ago, and the shape of their skull indicates these whales were capable of echolocation. Like modern whales, Squalodon had a thinned and hollowed out lower jaw, which would have accommodated fatty structures used to transmit sound echoes to the ear.*

Baleen whales evolved from toothed archaeocetes, and early baleen whale predecessors such as Llanocetus had both teeth and baleen. This partial skull, ear bones and skeleton date back 28 million years – possibly the earliest record for a right whale species.

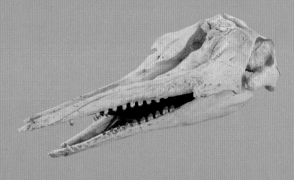

Waipatia maerewhenua *lived around 25 million years ago. The skull of this small toothed dolphin shows evidence of the ability to echolocate. Although now extinct, the species may be a distant ancestor of a present-day river dolphin.*

EWAN FORDYCE:

Uncovering whale fossils in Aotearoa New Zealand

I first became interested in fossils (and natural history in general) when I was about eight years old, prompted by the excitement of reading a book called *The World We Live In*. Years later, when I started university, I took geology as a topic to support my zoology degree and rediscovered fossils. I wanted to do research and started a doctoral degree in zoology and palaeontology, specifically looking at fossil whales and dolphins from New Zealand. It led me into a career that has proved stimulating and highly satisfying.

When I was a PhD student in the 1970s, I found several fossil penguins in rocks of South Canterbury, including remains of giant species belonging to the recently recognised (2012) genus of *Kairuku*. At the same time, I found several collections of fossil whale bones, including some that I dug out years later. Among the bones was a specimen that I recognised as a new species of 'dawn' baleen whales (eomysticetids); that specimen is now being studied by one of my PhD students. Early on, I also recovered the skull and jaws of a magnificent shark-toothed dolphin from rocks near Duntroon in North Otago.

Over the summer of 1986-87, I had a chance to go to Seymour Island in Antarctica with some American palaeontologists who were looking for bones of land mammals in shallow marine sedimentary rocks. Fossil whale bones had also been reported from the island and that was my focus. I spent several weeks looking for cetacean remains, without much luck. Then one day, the cloud rolled in and obscured the pathway back to the camp. I took an unexpected turn up a new gully and found in front of me scattered fossil bones, along with a block of rock containing two large teeth. More blocks of rock were identified as parts of a large skull. The material was shipped to New Zealand and assembled over about three years. The fragments were reconstructed as the 2-metre-long skull of a large whale that represented the transition between archaic toothed whales and modern baleen whales. The fossil, named *Llanocetus*, is the oldest reported from the baleen whale lineage. It lived about 35 million years ago, when

Antarctica was starting to cool dramatically and the south polar ice cap was forming.

Not long after the discovery of *Llanocetus*, I received funding from the National Geographic Society to search for fossil whales and dolphins in the South Island. One of the early finds from that work was a cluster of fossil dolphin teeth and ear bones from marine rocks in North Otago. It took several days to recover the finds from the field and about three months to prepare them. The fossil didn't fit into any group described formerly, so I named it *Waipatia maerewhenua* and placed it in a new family, Waipatiidae. *Waipatia* seems to have been an early dolphin in the lineage leading to the nearly extinct modern-day Ganges River dolphin, *Platanista gangetica*, and was probably similar in size to a bottlenose dolphin.

New Zealand rocks have provided other fossils that are significant in the wider evolutionary history of whales and dolphins: tusked squalodelphinids and dalpiazinids, which also seem to be distant relatives of *Platanista*; the impressive shark-toothed dolphins, or squalodontids; archaic ocean dolphins in the group Kentriodontidae; proto-rorquals; the distinctive local genus *Mauicetus*; a diversity of dawn baleen whales; an apparent ancient right whale (Balaenidae); and, surprisingly, a relict archaeocete whale that could not echolocate or filter-feed yet lived in the same seas as toothed dolphins and baleen whales.

One of my most satisfying recent papers deals with the relationships of the unusual pygmy right whale, *Caperea marginata*. The species has long been a taxonomic problem — it has either been allied closely with the right whales because of certain structural similarities, or with grey whales and rorquals on the basis of genetic similarities. Because of some specific skull characteristics, however, I wondered whether *Caperea* might actually be closer to a reportedly extinct group of archaic baleen whales called the cetotheres (Cetotheriidae), rather than to a particular living group of mysticete. One of my PhD students, Felix Marx, has a strong interest in relationships between baleen whales and compiled a large matrix of data on the structural characters of *Caperea*, other living baleen whales and many fossils, including cetotheres. His analysis showed that *Caperea* is indeed closer to cetotheres than to other groups of mysticetes. In 2012, we published our findings, in which we suggested that *Caperea* represents the last of the cetotheres, and that the family should therefore be resurrected from extinction.

UNDERSTANDING
WHALES

*Whales live in the midst of the sea like fish, yet
they breathe like land species … their watery realm extends
from the surface to the bottom most depths of the sea.*

Bernard Germain de Lacépède, *Histoire de Cétacés* (1804)

Currently, scientists recognise no fewer than 87 living (extant) cetacean species, which range in size from the diminutive 1.2-metre-long, 60-kilogram Hector's dolphin to the massive 33-metre-long, 150-tonne blue whale. The different species not only come in a wide range of sizes but also vary in their shape and colour: some are long and streamlined, while others are short and stocky; some have extraordinarily tall dorsal fins while others have no dorsal fin at all; and some are a dreary grey or brown, while others have striking colour patterns.

Habitat choices also vary, with some species preferring the tropics and others living in freezing polar conditions. Some live in deep offshore waters, while others prefer to be closer to shore, and a few species even spend most or all of their lives in estuaries or freshwater rivers.

New Zealand hosts a large percentage of the known cetaceans – at least 42 different species and three subspecies reside in, migrate past or are known to have inhabited our waters. A few of those species, however, such as the spectacled porpoise and some beaked whales, have never been sighted alive in New Zealand waters and are known here only from stranded specimens or a few bone fragments.

FASCINATING FACT

In 2011, DNA testing on the skeletal remains of female and juvenile whales that stranded the previous year at Opape Beach in the Bay of Plenty revealed that they were spade-toothed whales, *Mesoplodon traversii*, one of the rarest cetacean species in the world. The species was formerly known only from bone fragments discovered on the Chatham Islands in 1874, and skulls at White Island in the Bay of Plenty in the 1950s and in Chile in 1986.

The short-beaked common dolphin has a distinctive gold blaze running from behind its eye to its mid belly.

CETACEAN CLASSIFICATION

Living things are classified according to the Linnaean system, a binomial naming system devised by Swedish naturalist Carolus Linnaeus in the 18th century, in which every described species is given its own unique scientific name. Organisms are placed in categories according to the characteristics that they have in common, gradually being differentiated from one another down to species level through a descending hierarchy of ranks. Whales, for example, are placed in the kingdom Animalia (animals), phylum Chordata (vertebrates and their relatives), class Mammalia (mammals), superorder Laurasiathenia (placental mammals), order Cetartiodactyla (cetaceans and artiodactyls [even-toed ungulates]) and sub-order Cetacea (whales, dolphins and porpoises). The sub-order Cetacea contains three groups, Archaeoceti (ancient whales), Odontoceti (toothed whales) and Mysticeti (baleen whales) – and within each of these groups are various families, genera and species.

There are 13 extant cetacean families, four in the Mysticeti and nine in the Odontoceti, with members grouped according to shared characteristics. For example, in the family Phocoenidae, the porpoises, all seven members have spade-shaped teeth, while in the family Delphinidae, the dolphins, all 36 members have cone-shaped teeth.

The genus classification separates family members still further, with the genus name often identifying a specific characteristic. In the family Delphinidae, for example, both northern and southern right whale dolphins are placed in the genus *Lissodelphis*. The prefix *Lisso-* means 'smooth' and the suffix *-delphis* 'dolphin'; the combined name literally translates as smooth dolphin, reflecting the fact that both of these species do not have a dorsal fin on their back.

Each member within the genus is then given a species name, which is used in conjunction with the genus name. So, for example, the northern right whale dolphin is known as *Lissodelphis borealis* (literally 'northern smooth dolphin'), while the southern right whale dolphin is *Lissodelphis peronii* (named after French naturalist François Péron). Each cetacean species has a common name in addition to its binomial scientific name. Northern and southern right whale dolphins were given their common name because they resemble right whales in that they lack a dorsal fin.

HOW MANY?

The question of exactly how many cetacean species exist worldwide is a difficult one to answer. Scientific advances such as DNA testing have changed the way scientists acknowledge new species and subspecies. An example of this is the subspecies of Hector's dolphin known as Maui's dolphin, which lives off the North Island in New Zealand. Until recently all Hector's dolphins were classed as *Cephalorhynchus hectori*, but in 2002 DNA analysis confirmed that the dolphins living around the North Island were in fact genetically distinct enough to be classed as a separate subspecies, and they became known as Maui's dolphins (*Cephalorhynchus hectori maui*).

New cetacean species that have been recognised in recent years include the North Pacific right whale *Eubalaena japonica* (2000), once classed simply as a population of the northern right whale; Omura's whale *Balaenoptera omurai* (2003), once classed as Bryde's whale; and the newly described Perrin's beaked whale *Mesoplodon perrini* (2002).

While the number of cetacean species changes with new techniques and discoveries, there may also be conflicts as some scientific bodies accept the new species classification while others reject it. This is why there is an anomaly in numbers, with some scientists recognising 87 cetacean species worldwide and others accepting up to 90 species.

FASCINATING FACT

Carolus Linnaeus is famous for his 1735 publication *Systema Naturae*, in which he divided the natural world into animals, vegetables and minerals. Whales were classified as fish until the tenth edition came out in 1758 and he transferred them to the class Mammalia. When asked to comment on his work, Linnaeus would reply, 'God created, Linnaeus organised.'

THE ANCIENT WHALES (ARCHAEOCETI)

The primitive whales that lived during the Eocene epoch (56–34 million years ago) are known as the Archaeoceti. The vast majority of these early whales became extinct around 30–24 million years ago.

THE BALEEN WHALES (MYSTICETI)

No fewer than 14 baleen whale species exist worldwide, at least eight of which have been sighted in the waters around New Zealand. Instead of teeth, these whales have pieces of baleen (baleen plates) hanging from the roof of their mouth, which they use to filter food from the water. Baleen is made of keratin, the same substance that composes our hair and fingernails, but the plates are much thicker. Each plate has frayed, hairy ends, which intertwine to form a fibrous mat that acts like a colander and strains prey from the water.

Baleen is often referred to as whalebone and was once used in making products that require flexibility, such as upholstery and women's corsets. Some mysticetes have more than 600 pieces of baleen in their mouth, these ranging in length from 30 centimetres to more than 3 metres. All baleen whales breathe through two blowholes, and females are usually larger than males.

THE TOOTHED WHALES (ODONTOCETI)

Worldwide, there are 73 extant toothed cetacean species, of which no fewer than 34 are known from New Zealand waters. Their teeth range in size and number, with some species having only one tooth (or, in some female beaked whales, no erupted teeth at all) and others having over 250. All toothed cetaceans echolocate, using sound to build up a picture of what is around them (see page 50). They also possess a single external blowhole (baleen whales have two blowholes), and males are usually larger than females.

Narwhals lack functional teeth and feed by sucking in their prey. In males, the left-hand upper jaw tooth erupts through the lip and grows into a long, spiralled tusk, reaching up to 2.7 metres in length.

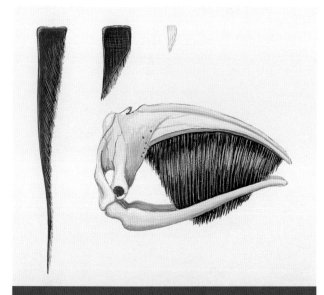

Right whale baleen is dark brown to dark grey or black, up to 2.8 metres long and dense but with fine, fringed ends (left). Humpback whale baleen is black to dark brown, 70–100 centimetres long and coarse (centre). The fine bristles of minke whale baleen is creamy white to yellow and 20–30 centimetres long (right).

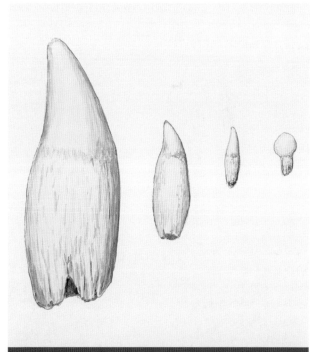

Teeth from a sperm whale, orca, dolphin and porpoise (left to right).

CETACEAN ORGANISATION

———

The evidence that artiodactyls, even-toed ungulates such as cattle, camels, pigs and hippos, are the closest living relatives of whales led to the creation of a new rank with the hybridised name Cetariodactyla for the two groups. The name is a combination of the Latin *cetus* ('large sea creature'), and the Greek *artios* ('even numbers') and *daktulos* ('finger' or 'toe'). There has been debate as to whether Cetariodactyla should be a superorder or just the order for whales; current thinking places whales in the order Cetariodactyla, with Cetacea becoming the sub-order. The species list below is cited from the 'List of Marine Mammal species and subspecies' (Committee on Taxonomy, Society for Marine Mammalogy, www.marinemammalscience.org, 2012). The list is expected to be revisited and revised at regular intervals, reflecting the continuing flux in marine mammal taxonomy. It currently lists 88 cetacean species – 87 extant and one extinct (baiji, Yangtze River dolphin). Species known in New Zealand are marked with *.

ORDER: CETARIODACTYLA

SUB-ORDER: CETACEA

ODONTOCETI (TOOTHED WHALES)

Family PHYSETERIDAE

The family Physeteridae contains a single species, the sperm whale, which is the largest living toothed predator on Earth.

Genus PHYSETER
Sperm whale *Physeter macrocephalus**

Family KOGIIDAE

Genus KOGIA
Pygmy sperm whale *Kogia breviceps**
Dwarf sperm whale *Kogia sima**

Family DELPHINIDAE

Confusingly, orca, the largest dolphin, is also known as the killer whale, and five other species in the Delphinidae family are also called whales.

Genus CEPHALORHYNCHUS
Hector's dolphin
*Cephalorhynchus hectori**
Commerson's dolphin
*Cephalorhynchus commersonii**
Chilean dolphin *Cephalorhynchus eutropia*
Heaviside's dolphin
Cephalorhynchus heavisidii

Genus DELPHINUS
Short-beaked common dolphin
*Delphinus delphis**

Long-beaked common dolphin
Delphinus capensis

Genus LAGENORHYNCHUS
Dusky dolphin *Lagenorhynchus obscurus**
Hourglass dolphin
*Lagenorhynchus cruciger**
Peale's dolphin *Lagenorhynchus australis*
Atlantic white-sided dolphin
Lagenorhynchus acutus
White-beaked dolphin
Lagenorhynchus albirostris
Pacific white-sided dolphin
Lagenorhynchus obliquidens

Genus LAGENODELPHIS
Fraser's dolphin
*Lagenodelphis hosei**

Genus LISSODELPHIS
Southern right whale dolphin
*Lissodelphis peronii**
Northern right whale dolphin
Lissodelphis borealis

Genus STENELLA
Striped dolphin *Stenella coeruleoalba**
Pantropical spotted dolphin
*Stenella attenuata**
Spinner dolphin *Stenella longirostris*
Clymene dolphin *Stenella clymene*
Atlantic spotted dolphin *Stenella frontalis*

Genus STENO
Rough-toothed dolphin *Steno bredanensis**

Genus TURSIOPS
Common bottlenose dolphin
*Tursiops truncatus**

Indo-Pacific bottlenose dolphin
Tursiops aduncus

Genus GRAMPUS
Risso's dolphin *Grampus griseus**

Genus SOUSA
Pacific humpback dolphin
Sousa chinensis
Atlantic humpback dolphin
Sousa teuszii

Genus SOTALIA
Tucuxi dolphin *Sotalia fluviatilis*
Guiana dolphin *Sotalia guianensis*

Genus ORCAELLA
Irrawaddy dolphin
Orcaella brevirostris
Australian snubfin dolphin
Orcaella heinsohni

Genus ORCINUS
Killer whale (orca) *Orcinus orca**

Genus GLOBICEPHALA
Long-finned pilot whale
*Globicephala melas**
Short-finned pilot whale
*Globicephala macrorhynchus**

Genus PSEUDORCA
False killer whale
*Pseudorca crassidens**

Genus FERESA
Pygmy killer whale
*Feresa attenuata**

Genus PEPONOCEPHALA
Melon-headed whale
*Peponocephala electra**

Family MONODONTIDAE

Belugas and narwhals live only in the far north of the northern hemisphere.

Genus DELPHINAPTERUS
Beluga *Delphinapterus leucas*

Genus MONODON
Narwhal *Monodon monoceros*

Family PLATANISTIDAE

Genus PLATANISTA
Indian river dolphin *Platanista gangetica*

Family INIIDAE

Genus INIA
Amazon River dolphin (boto) *Inia geoffrensis*

Family PONTOPORIIDAE

Genus PONTOPORIA
La Plata River dolphin (Franciscana) *Pontoporia blainvillei*

Family LIPOTIDAE (Extinct)

The Yangtze River dolphin (baiji) was last sighted in 2002. Now declared extinct, it is the first known cetacean species to be exterminated by human activity.

Genus LIPOTES
Yangtze River dolphin (baiji) *Lipotes vexillifer*

Family ZIPHIIDAE

The beaked whales, Ziphiidae, are the second largest cetacean family and one of the most diverse and mysterious. Some species have never been sighted alive and are known only from stranded remains.

Genus ZIPHIUS
Cuvier's beaked whale *Ziphius cavirostris**

Genus TASMACETUS
Shepherd's beaked whale (Tasman beaked whale) *Tasmacetus shepherdi**

Genus BERARDIUS
Arnoux's beaked whale *Berardius arnuxii**
Baird's beaked whale *Berardius bairdii*

Genus HYPEROODON
Southern bottlenose whale *Hyperoodon planifrons**
Northern bottlenose whale *Hyperoodon ampullatus*

Genus INDOPACETUS
Indo-Pacific beaked whale (Longman's beaked whale) *Indopacetus pacificus*

Genus MESOPLODON
Hector's beaked whale *Mesoplodon hectori**
True's beaked whale *Mesoplodon mirus**
Gray's beaked whale (Scamperdown whale) *Mesoplodon grayi**
Andrews' beaked whale *Mesoplodon bowdoini**
Blainville's beaked whale (dense-beaked whale) *Mesoplodon densirostris**
Hubb's beaked whale *Mesoplodon carlhubbsi*
Sowerby's beaked whale *Mesoplodon bidens*
Gervais' beaked whale *Mesoplodon europaeus*
Layard's beaked whale (strap-toothed whale) *Mesoplodon layardii**
Ginkgo-toothed beaked whale *Mesoplodon ginkgodens**
Perrin's beaked whale *Mesoplodon perrini*
Stejneger's beaked whale *Mesoplodon stejnegeri*
Pygmy beaked whale *Mesoplodon peruvianus**
Spade-toothed whale *Mesoplodon traversii**

Family PHOCOENIDAE

The sole porpoise found in New Zealand, the spectacled porpoise, is known here only from the remains of a few stranded animals.

Genus PHOCOENA
Spectacled porpoise *Phocoena dioptrica**
Harbour porpoise *Phocoena phocoena*
Burmeister's porpoise *Phocoena spinipinnis*
Vaquita *Phocoena sinus*

Genus PHOCOENOIDES
Dall's porpoise *Phocoenoides dalli*

Genus NEOPHOCAENA
Indo-Pacific finless porpoise *Neophocaena phocaenoides*
Narrow-ridged finless porpoise *Neophocaena asiaeorientalis*

MYSTICETI (BALEEN WHALES)

Family BALAENOPTERIDAE

All species within the family Balaenopteridae have expandable throat pleats; they are commonly known as the rorquals, from a Danish word meaning 'tubed' or 'pleated'.

Genus BALAENOPTERA
Common minke whale *Balaenoptera acutorostrata**
Antarctic minke whale *Balaenoptera bonaerensis**
Bryde's whale *Balaenoptera edeni**
Omura's whale *Balaenoptera omurai*
Sei whale *Balaenoptera borealis**
Fin whale *Balaenoptera physalus**
Blue whale *Balaenoptera musculus**

Genus MEGAPTERA
Humpback whale *Megaptera novaeangliae**

Family BALAENIDAE

Right whales and bowheads lack a dorsal fin and do not have throat pleats. The southern right whale is the only Balaenidae species found in the southern hemisphere.

Genus EUBALAENA
North Atlantic right whale *Eubalaena glacialis*
North Pacific right whale *Eubalaena japonica*
Southern right whale *Eubalaena australis**

Genus BALAENA
Bowhead whale *Balaena mysticetus*

Family NEOBALAENIDAE (soon to be CETOTHERIIDAE)

Pygmy right whales, the smallest baleen species, are seldom sighted and little is known about their lifestyle or habits. A 2012 scientific paper gives evidence that pygmy right whales should be moved to the family Cetotheriidae (previously thought to be extinct), in which case Neobalaenidae will become defunct.

Genus CAPEREA
Pygmy right whale *Caperea marginata**

Family ESCHRICHTIIDAE

The family Eschrichtiidae contains a single species, the grey whale, which is found only in the northern hemisphere.

Genus ESCHRICHTIUS
Grey whale *Eschrichtius robustus*

WHALE, DOLPHIN OR PORPOISE?

The generic term 'whales' is often used when referring to all the cetacean species. This leads to some confusion over the terms 'whale', 'dolphin' and 'porpoise'. Basically:

'Whale' normally refers to the larger toothed and baleen species.

'Dolphin' refers to medium-sized cetaceans that usually have a beak and always have conical teeth.

'Porpoise' refers to smaller-sized cetaceans that lack a beak and have flat, spade-shaped teeth.

But wait, there are anomalies:

- Six species that have 'whale' as part of their common name, including the pilot whale and killer whale, are actually members of the dolphin family.

- Some whales, such as the pygmy right whale, are smaller than large dolphins such as the common bottlenose.

- Some dolphins, such as dusky and Hector's, have no beak, and the tiny Hector's is actually smaller than most porpoises.

Killer whales are the largest species in the Delphinidae family.

People have often referred to Hector's dolphins as 'porpoises' because of their small size and lack of beak.

FORM AND FUNCTION

A cetacean's body is adapted for life in the sea. Its flippers, dorsal fin and tail flukes are the only projections along an otherwise streamlined form. The almost perfect hydrodynamic teardrop body shape enables these animals to swim quickly and gracefully with a minimum of drag. A dolphin's skin feels soft and smooth but it is actually covered with minute microfolds that allow water to flow past smoothly when they are swimming through choppy seas.

Beneath the surface of a cetacean's skin is a layer of insulating blubber that helps them maintain a body temperature of 37°C. Whales that migrate from the poles to the tropics actually need to cool down in warmer waters. They accomplish this by increasing blood flow to their extremities (flippers, tails and dorsal fins), thereby radiating heat from their blood into the relatively cooler waters surrounding them. Blubber is important in other ways: it helps cetaceans maintain buoyancy and also acts as a rich energy store for baleen whales that are fasting or consuming very little food in their breeding grounds. The thickness of blubber ranges from a few centimetres in dolphins to 60 centimetres in right and bowhead whales.

The mammary glands, genitalia and excretory organs of cetaceans are concealed within the body in order to reduce drag when the animal is swimming. Even the external ear has been replaced with a tiny pinhole located just behind each eye.

Whale blubber is a combination of fat cells surrounded and held in place by a structural mesh of collagen fibres. The blubber also contains numerous blood vessels and specialised shunts to allow efficient circulation of blood, thereby aiding thermoregulation. Here 'blanket pieces' of blubber are being cut on a factory ship during the 1930s.

FLIPPERS AND FINS

The bones in cetacean flippers are similar to those in a human hand, the main difference being that the human hand has mobile joints while the whale's flipper is rigid, somewhat like a paddle. The flippers give stability and are also used to help the animals steer and turn as they swim.

A cetacean's dorsal fin and tail flukes do not have bones; instead, they are composed of a strong muscular material. The powerful flukes are positioned horizontally, thus propelling the animal forward with every up and down stroke of the tail.

Socially, cetaceans use their flippers, dorsal fins and tails in a variety of ways. During courtship, they will often use their flippers to stroke each other, and they sometimes swim close together with their flippers touching. After giving birth, a mother may support her new calf with her dorsal fin, and sick or injured animals may be supported at the surface in the same way. Tail flukes, dorsal fins and flippers are also used in play, such as when a whale picks up and flicks off a piece of weed, or passes it on to another in the vicinity. When a flipper or tail is slapped on the water, it creates a loud cracking sound, which is thought to be used as a form of communication and to assist in herding prey.

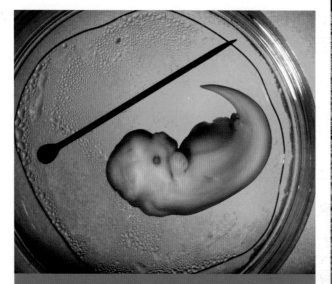

This dolphin embryo from the Thewissen Lab at the Northeast Ohio Medical University illustrates the similarity between foetal cetaceans and human foetuses. Initially, a cetacean embryo has two front flipper buds and two tiny hind limb buds where the arms and legs in a human embryo are. As the embryo develops, the rear buds disappear and a small tail starts to grow.

Dorsal fins come in a variety of shapes and sizes. It is thought that they may act as a stabiliser, although a number of species manage very well with only a tiny fin or small hump, and some species, such as the southern right whale (top), have no dorsal fin at all. Male orca (bottom) have the tallest dorsal fin, reaching a height of 1.8 metres, the size of a grown man.

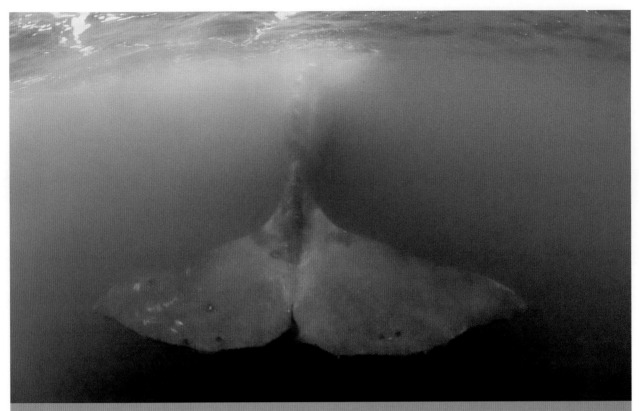
Tail flukes vary between species. They may be V-shaped, straight or concave along their trailing edge.

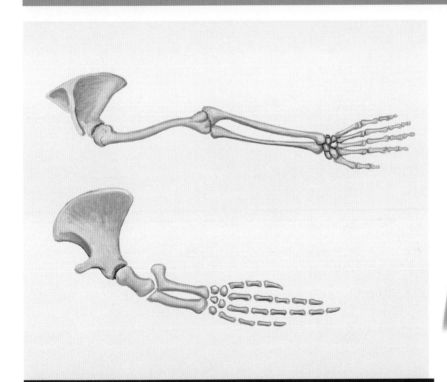
A whale flipper (bottom) contains bones that are similar to a human hand (top). The cetacean bones are enclosed within a common skin surface and there are no remaining vestiges of nails on the 'finger' bones.

FASCINATING FACT

The humpback whale's genus name, *Megaptera*, means 'big wing' and refers to the animal's long flippers, which reach lengths of 5 metres or more and are one of the longest appendages in the animal kingdom.

BREATHING AND BLOWING

As whales evolved into fully aquatic mammals, the nose gradually migrated to the top of the head, which enabled them to breathe more efficiently when swimming through the sea. In time, the external nostrils disappeared completely, to be replaced by one or two blowholes. Whales cannot breathe through their mouth, so their only form of respiration is through the blowhole.

In order to breathe, whales must consciously inhale and exhale; it is not an automatic process, as it is in humans. When whales become unconscious, their blowholes automatically close and they do not breathe. Therefore, since whales are conscious breathers, they are unable to sleep as we do. In order to rest, they shut down one side of their brain at a time and normally lie very quietly on or just below the surface of the water.

Before a whale dives, a powerful set of muscles closes the blowhole, making it watertight.

Most blowholes are on the top of the head but in sperm whales they are forward on the left-hand side, so the blow is angled forward and to the left.

'Thar she blows' was the call that went up when whalers sighted a breathing whale in the distance. Visually, a breathing whale appears to be exhaling water but this is not the case. The spout is actually water vapour, created when the warm, moist air in the whale's lungs is exhaled and condenses in the relatively cooler air outside. A whale's fishy-smelling 'blow' also consists of mucus, along with any water that has been trapped around the blowhole.

FASCINATING FACT

Whales may exchange 80–90 per cent of the air in their lungs with each breath; in comparison, humans exchange only about 10 per cent.

Toothed whales have one external nostril or blowhole, (common dolphin, top) while baleen whales have two.

The shape of some blows can be used to identify whale species: right whales have a V-shaped blow (top); in humpbacks, it's bushy, pear-shaped (centre); blue whales have the tallest blow of any whale, up to 6 metres high (bottom).

BRAIN POWER

On the planet Earth, man had always assumed that he was more intelligent than dolphins because he had achieved so much – the wheel, New York, wars and so on – whilst all the dolphins had ever done was muck around in the water having a good time. But conversely, the dolphins had always believed that they were far more intelligent than man – for precisely the same reasons.

Douglas Adams, *The Hitchhiker's Guide to the Galaxy* (1979)

The question of cetacean intelligence is often asked but very difficult to answer. Obviously we cannot test the IQ of a whale, but we can look at relative brain size to body weight, which is one indicator of so-called intelligence. In general, the more predatory toothed whales have a larger brain size in relation to their body weight than the grazing baleen whales; in fact, some dolphin species such as the bottlenose have a relative brain size (around 1 per cent) close to that of humans (around 2.3 per cent). Toothed cetaceans also have a highly developed frontal cortex, again supposedly a sign of higher intelligence. Studies carried out on captive cetaceans, particularly bottlenose and rough-toothed dolphins, indicate an ability for complex and abstract cognition.

It is hard, however, to speculate on the intelligence of whales and dolphins, which naturally occur in a liquid environment that is totally foreign to our own. We do know that the toothed species have developed sophisticated skills such as echolocation and that many species have diverse feeding techniques that require extensive communication, coordinated hunting strategies and an awareness of their prey's behaviour.

As well, most cetaceans live in complex social societies that nurture and teach their young. Research indicates that, in a similar way to humans, behaviours and skills are communicated and passed on through generations, resulting in distinct cultures in some cetacean societies, such as humpback and sperm whales, and orca and other dolphin species. It is possible that other species, not as well studied, also display distinctive cultural traits. In addition, many cetaceans 'play' – not just the young, as occurs in most mammal species, but adults as well. Examples include dolphins carrying bits of seaweed and flicking them back and forth to each other, large whales playing with objects in the water for hours at a time, and dolphins pulling on the legs of resting seabirds – in this last case, the animals seem to be displaying a sense of humour, supposedly a sign of higher intelligence.

In this feeding session, a group of around 12 dusky dolphins worked together to prey on kahawai. A few of the dolphins encircled the fish school to keep it in a tight ball, while the others dove underneath to pluck out the fish of their choice. During the whole activity, the duskies communicated continuously, emitting a combination of echolocation clicks alongside squeaks and squeals.

All rorquals have expandable throat pleats (some species have more than 100), so they can literally take in tonnes of prey and water in just one gulp.

FOOD AND FEEDING

Baleen and toothed whales catch their food in different ways. Baleen whales feed on minute zooplankton, such as vast swarms of krill or copepods, or small schooling fish species, which they filter from the water through their plates of baleen. As the baleen species evolved, they developed feeding techniques that enabled them to consume large quantities of prey in a single mouthful. In contrast, toothed whales catch their prey (normally squid or fish) one at a time. Finding and catching individual prey requires different skills, one of which is a highly developed sonar system known as echolocation (see page 50).

BALEEN WHALES: GULPERS, GRUBBERS AND SKIMMERS

Mysticetes are batch feeders, filtering prey through the frayed ends of their baleen, and use one of three feeding techniques. The gulpers, such as humpback, minke, Bryde's, sei, fin and blue whales, possess expandable pleated throat grooves and are collectively known as rorquals; the grubbers, the grey whales, have short, coarse baleen; and the skimmers, such as right and bowhead whales, have enormous mouths and long baleen.

Gulpers

Rorquals feed by lunging through congregations of prey, gulping in huge mouthfuls of the 'seafood broth'. The whales push the water out of their mouth with their large, fat tongues and then swallow the prey left tangled in their baleen trap. Blue whales have finely fringed baleen and feed mainly on zooplankton such as krill, while many other rorqual species have coarser baleen and feed on both zooplankton and small schooling fish.

Humpbacks have the most varied and spectacular feeding techniques of the gulpers. In addition to normal

FASCINATING FACT

During their summer feeding season at the poles, blue whales will consume more than 4 tonnes of krill in a single day, equivalent to about 4 million individual crustaceans.

lunge-feeding, they have also been observed stunning their prey by slapping the water with their large flippers or tails. Their most impressive feeding technique, however, is bubble netting. In this, a group of whales swims in a slow circle, blowing out bubbles to form a 'net' around their prey. The krill or schooling fish group closer together as the bubbles encircle them, and then the whales lunge up through the centre of the net, gulping in vast quantities of food. Sometimes the bubbles are tiny, at other times they're the size of a large pizza, while some of the bubble nets themselves are large enough to enclose a four-storey house.

Researchers using underwater video cameras have recently discovered that bubble netting may be more complex than first thought. They found that the whales in their study group have specialist roles, some making hunting sounds that help herd prey, while one or two others swim down to create the bubble net. It also appears that certain individuals form feeding teams, which work together from year to year. In essence, the whales are employing cooperative long-term relationships between non-related individuals and using tools (bubbles) to capture their prey.

This skim-feeding right whale is skimming at the water's surface, which allows the water to flow out the sides of its mouth leaving the whale's prey trapped in its baleen.

Grubbers

Grey whales, found only in the northern hemisphere, are primarily bottom feeders that grub around on the ocean floor, stirring up sediments containing prey such as sea worms, small crustaceans and tiny fish. Once the sediments have been stirred up, the whales turn on their side and suck in the mix, filtering out prey and expelling the remainder. The whales also use gulping and skimming techniques to feed on mid-water and surface species.

Skimmers

Bowhead and right whales feed by swimming with their mouths open, either on or just under the water's surface. Inside the huge mouth are up to 600 pieces of baleen more than 2 metres long. Each piece of baleen has fine fringed hair on its inside edge, creating the perfect trap for the whales' main prey, copepods. The versatile sei whale is generally a gulper but will also skim-feed on occasion – the only rorqual that does so.

Humpback whales use a gulping technique to feed, lunging through shoals of krill or schooling fish.

TOOTHED WHALES: GRASPERS AND SUCKERS

Although they have teeth, odontocetes do not chew their food but instead use them to grasp it (the 'graspers') or suck it into their mouth (the 'suckers') and then swallow it whole. The size and number of teeth vary between species: dolphins may have more than 250 teeth, while most male beaked whales have just one to two oddly shaped teeth and the female beaked whales usually have none at all. Sperm whales have 20–24 pairs of large, impressive-looking teeth in their lower jaw, yet they are probably seldom used since the whales suck up the majority of their food.

Most toothed cetaceans consume a mixed diet of fish and squid, but many species do have a preference for one or the other. As a general rule, odontocetes that feed primarily on fish, such as dolphins and porpoises, tend to be graspers, while those that feed primarily on squid, such as sperm and beaked whales, tend to be suckers. Killer whales have the most varied diet, which ranges from numerous fish species – including rays and sharks – to other marine mammals such as seals, dolphins and even large whales.

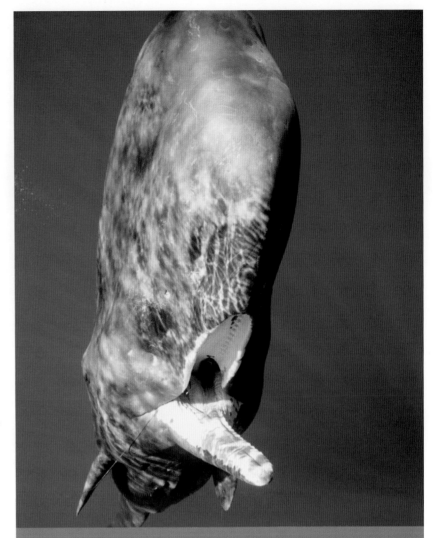

Sperm whales mainly use suction to capture their food although large prey such as giant squid may be grabbed first and then sucked in and swallowed.

This juvenile humpback whale has had a lucky escape. Its entire body is covered with tooth rake marks after an attack by a pod of orca, which also bit off the tip of the whale's dorsal fin and took large chunks out of its tail flukes.

SEEING WITH SOUND

Moving through a dim, dark, cool watery world of its own, the whale is timeless and ancient... prowling the ocean floor a half-mile down, under the guidance of powers and senses we are only beginning to grasp.

Victor B. Scheffer, *The Year of the Whale* (1969)

Baleen whales feed near the surface where there is light to see, but many toothed species feed in murky waters or at depths where the light is very dim or non-existent. Because they can't navigate or locate prey in these dark environments using sight, the odontocetes have instead evolved the ability to echolocate, essentially seeing with sound.

The toothed whales echolocate by emitting bursts of broad-spectrum pulses or clicks that travel through the water as sound waves. When these hit a solid object they bounce back to the whales, which then interpret the returning echoes. Each object produces a different sound echo, and as sound waves in water travel at about 1500

metres per second – over four times faster than in air – echolocation is extremely quick and efficient. Experiments with dolphins have shown that not only can they locate tiny objects from a great distance, they can also determine the size, shape and density of the objects. Some objects can be analysed with just a few clicks, but when the whale or dolphin is searching for prey, analysing it and then chasing it down, thousands of clicks may be required.

Most scientists believe that the initial sound waves are created by the cetacean when air is forced out through a pair of fat, gristly flaps in the nasal tubes, known as phonic lips. As the 'lips' vibrate, the sound is transmitted through the whale's forehead, known as the 'melon', which is composed of fatty tissue with the same acoustic properties as water. The melon is thought to work like a lens, focusing the sound waves forward and into a beam. Toothed whales may change the direction of the sound by using facial muscles to alter the shape of the melon or by bouncing the sound off air sacs located below the blowhole. The animals

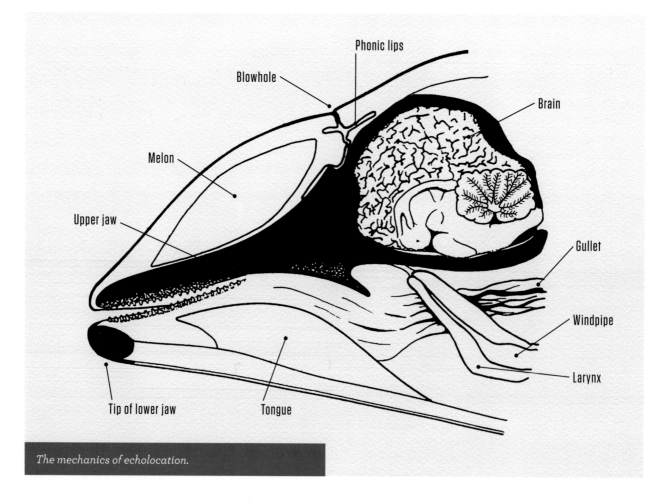

The mechanics of echolocation.

send out high-frequency clicks to gain a detailed picture of objects at close range, while low-frequency clicks are used for larger objects and for greater distances.

The returning echoes are believed to be collected by a deposit of fat in the animal's lower jaw. The sound waves are then relayed through a thin channel to the ear bones and transmitted to the brain for processing and interpretation.

While sperm and beaked whales are generally solitary hunters that use echolocation to locate and capture prey, most dolphins and porpoises hunt together in a group using echolocation along with a variety of other communicative sounds.

HUNTING USING ECHOLOCATION

As the sperm whale dives, it closes its blowhole, lowers its head and lifts its powerful tail flukes. The blowhole remains closed for the duration of the dive, which lasts on average 30–45 minutes but can be as long as two hours.

The descent speed is a steady 3–6 kilometres an hour. Right from the onset, the whale starts emitting a series of slow, evenly spaced echolocation clicks, guiding it on its underwater quest for food.

As the whale moves into the mesopelagic zone (see page 12), light has all but disappeared. The increasing pressure causes physiological changes: the whale's flexible ribcage partially caves in as its lungs start to collapse, forcing remaining air into the nasal passages. Myoglobin-rich muscle tissue stores over 40 per cent of the animal's total oxygen supply used on the dive. To conserve oxygen, the heart rate slows and oxygen-rich blood is diverted to vital organs such as the heart and brain.

Much of the sperm whale's prey lives in the mesopelagic zone, especially the mid-water squid species that make up the majority of its diet. Echolocation clicks remain evenly spaced until prey is located, at which point they get increasingly faster as the whale closes in, ultimately running together so that individual clicks become indistinguishable – a so-called click train. Most of the squid species are relatively inactive and bioluminescent, so are easy to catch, but a few require active hunting. Some scientists theorise that prey may be attracted to the white area surrounding the whale's lower jaw, or that the echolocation bursts may stun the prey, but neither of these theories has been proven.

Sperm whales sometimes continue down into the bathypelagic zone (see page 13), where the total darkness is broken only by flashes of bioluminescent light emitted by its residents, including the giant squid. Deep scratches on sperm whales made by the beaks of these squid indicate that this prey is not always an easy meal. The whales have been tracked to depths of 2800 metres but may descend to the ocean floor as the remains of bottom-dwelling sharks that live at 3000 metres have been found in sperm whale stomachs.

Using evenly spaced echolocation clicks as a guide, the whale begins its ascent to the surface. There it rests and re-oxygenates for at least 10 minutes, before beginning another descent into the deeps.

A sperm whale normally begins to echolocate as soon as its tail disappears beneath the surface of the sea.

CETACEAN COMMUNICATION

Whales and dolphins communicate with each other in a variety of ways. Some communication is physical, such as touching, stroking or overt behaviour like breaching or lob-tailing. Cetaceans also communicate using a wide repertoire of vocalisations and different sound frequencies.

COMMUNICATION AMONG ODONTOCETES

Most toothed whales and dolphins communicate using echolocation combined with a variety of other sounds that include squeaks, whistles, chirps, yelps and squawks. Interestingly, however, the largest toothed species, the sperm whale, and the smallest, the Hector's dolphin, use only echolocation clicks for communication and for locating prey.

Sperm whales

When a sperm whale begins its search for food, it uses slow, evenly spaced echolocation clicks, these gradually becoming faster and closer together as the whale approaches its prey, forming a 'click train'. The clicks of an individual whale sound like a person hammering in a nail, starting off with a few evenly spaced blows and then getting quicker as the nail is driven into place, and a group of feeding sperm whales sound like a gang of carpenters on a construction site. The clicks of socialising sperm whales have a very different tone and cadence. These clicks, or codas, are softer and more uneven, a bit like the dots and dashes of Morse code.

Researchers studying South Pacific sperm whales, such as Hal Whitehead of Canada's Dalhousie University, have discovered that the animals' codas have a unique rhythmic pattern. Different social groups or units of whales use their own distinctive codas, and members of these groups belong to separate clans that share similar distribution patterns and behaviours. The researchers have identified four to five different clans in the South Pacific. These clans are matrilineal, with mothers passing on their clan's distinctive coda patterns to their offspring.

Hector's dolphins

New Zealand's Hector's dolphins are commonly found in murky water with limited visibility. The dolphins use extremely fast, high-frequency echolocation clicks averaging 120 kilohertz, about six times higher than the range of human hearing. The clicks are emitted in sequences ranging from dozens to thousands depending on the difficulty

In clear water, whales or dolphins may communicate by using visual clues but in murky or dark water they must rely on verbal sounds or echolocation for communication.

of the task at hand. An easy task, such as locating other dolphins, may require only a few dozen clicks, whereas a more complicated task, such as finding and chasing down prey, may require thousands of clicks. Stephen Dawson, a researcher into Hector's dolphins at Otago University, once recorded an amazing rate of over 1100 clicks in a single second.

Killer whales

Killer whales, or orca, communicate with a wide variety of calls, including echolocation clicks, pulsed calls and high-pitched squeals and whistles. Studies in the USA's Pacific Northwest, in British Columbia and in Norway have shown that members of resident killer whale populations living in those areas have a universal vocabulary of common calls, but that separate family pods within these general populations have their own distinct dialects. In essence, each pod has a different 'pronunciation' and sometimes even a slightly different usage of the common calls. Many of the orca live in the same matrilineal pod throughout their lives, but sometimes a sub-group will form that ultimately becomes its own separate pod. When this occurs, the new pod initially retains its original dialect, but subtle variations occur over time and ultimately the new pod develops its own unique dialect. Pods with related dialects belong to a clan. Scientists believe it is likely that future studies of killer whale populations in different parts of the world will reveal similarities in their use of calls.

COMMUNICATION AMONG MYSTICETES

Baleen whales communicate mainly by emitting moans, thumps, chirps and whistles, although the right whale's noisy calls also include growls, shrieks, chortles, snarls,

Not all communication is verbal. Breaching may be used to announce a whale's presence or as a sign of intimidation or aggression.

Killer whale populations that feed mainly on a variety of fish species are known to use a range of sounds, whereas those that feed primarily on other marine mammals tend to be silent hunters, using very reduced sounds or no sound at all when they are hunting their prey.

A RECORD OF A SONG

Humpbacks became an iconic symbol for environmentalists in the 1960s and 1970s. In 1970, a booklet entitled *Save the Whales* was published along with a record of humpback whales singing in the ocean depths. The recording, *Songs of the Humpback Whale*, went on to sell more than 30 million copies. In 1979, the recording was included as a sound page in *National Geographic* magazine. Ten million copies were printed and sent around the world – the largest single pressing of any record in history.

whoops, howls, sighs, roars, yaps, yelps, yips, barks and brays. Most of these sounds are low frequency. Blue and fin whales produce the lowest frequencies of all, their sounds taking the form of moans that can last up to half a minute and are usually repeated at regular intervals of 70–140 seconds. The moans of both species are not only the lowest but also the loudest sounds produced by any animal, averaging 155–188 decibels – equivalent to a jet engine a metre away running at full power!

Low-frequency sounds can travel great distances – Cornell University's Chris Clark once used sophisticated US Navy equipment to track a blue whale from as far away as 1600 kilometres over a period of 43 days. The whale travelled almost 3500 kilometres during that time, and according to Clark 'sang' continuously day and night. While the purpose of that extended song remains a mystery, some scientists have suggested that the low frequency calls of both blue and fin whales may enable them to maintain contact with one another across entire oceans.

Humpback whales

Early whalers have noted that humpbacks made 'peculiar cries', but for years nobody really paid much attention to their sounds. In fact, it could be said that international political intrigue led to the discovery of their songs.

In the early 1950s, when the Soviet Union and USA were engaged in the Cold War, each country was paranoid about the activities of the other and a great deal of spying took place. At the time, Frank Watlington of the US Navy was given the task of listening for Russian submarines off Bermuda and Hawai'i. In the first few years, Watlington noted the occurrence of 'strange sounds from an unknown source', but by the mid-1950s he realised the mysterious noises were being made by humpback whales. The recordings he made were considered classified information, so it wasn't until 1967 that the humpback sounds were finally released to whale researchers Roger Payne and Scott McVay. When the pair analysed the recordings, they realised the sounds had a regular rhythm and were full of patterns and organisation. After further analysis, they came to the astonishing conclusion that the humpback whales were, in essence, singing songs.

Payne and McVay discovered that each humpback song is composed of different sound units that are combined into recurring phrases, and that these are then repeated several times to make up a theme. Most songs are made up of five to seven themes that are repeated in a sequential order. A song lasts 5–30 minutes and may sometimes be repeated over and over in a song session lasting for several hours.

For years it was thought that humpback songs were part of a courtship ritual designed to attract females. However, researchers have recently discovered that there is little outward female response to singing males and that the songs mainly lead to social interactions between adult males. It appears that normally, when a male is singing, one of two events occurs:

1. The singing male is approached by another male for a short period of time. The approaching, non-singing male may simply pass by the singing whale, or one or both whales may engage in active behaviours such as lob-tailing or breaching. The whales then separate and

the first whale resumes singing and, at times, the other whale also begins to sing.

2. The singing whale may stop vocalising and join a passing group of whales that includes a potentially breeding female.

The exact reasons why humpbacks sing are still unknown. What is known is that the song is a communication from males that usually takes place on breeding grounds, and that it broadcasts the location of the singer or singers and may announce their ability or willingness to breed.

We have learned a great deal about the songs themselves. Whales in different oceans sing different songs, whales in the same ocean sing similar songs, and all of the whales on the same breeding ground sing the same song.

Even though all the males on a particular breeding ground sing the same song at the beginning of a season, that song does not remain the same. At some point, one of the whales makes a change to the tune, and within days all of the other males have copied the change. The song continues to be altered as the mating season progresses, so that by its end the whales are singing the new version. The following year, the males arrive singing the 'new' song from the previous season, which will again change as the current season progresses. Researchers have analysed humpback songs for more than 40 years and have never heard the same song repeated. (For more on humpback song research, see page 150.)

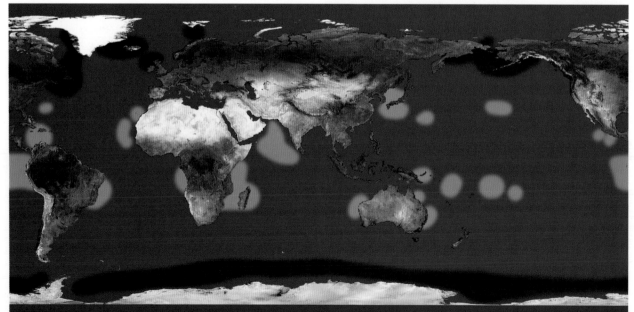

Humpback breeding grounds are located at latitudes of around 20° in the northern and southern hemispheres (light blue) and their feeding grounds are near the poles (dark blue).

'It has been known for a long while that humpback-whales ... have their own peculiar cry, or as whalemen express it, "sing".'

H.L. Aldrich, Arctic Alaska and Siberia, or, Eight Months with the Arctic Whalemen *(1889)*.

HUMPBACK SONG FACTS

- Only male humpbacks sing songs.

- Some singing occurs at the end of the summer feeding season and during migration, but most songs are sung on the winter breeding grounds.

- A singing whale often hangs in a head-down, tail-up position about 15–30 metres below the surface. Songs average 8–15 minutes long but sometimes last up to 30 minutes, at which point the whale surfaces to

breathe and then resumes its underwater singing position.

- Humpbacks sing one of the longest songs in the animal kingdom.

- While many singers remain stationary, others travel tens or even hundreds of kilometres during a song session and may even be accompanied by a female and calf.

LIVING TOGETHER

Different species of cetaceans have adopted different ways of living with each other. Some spend much of their time alone, while others roam the ocean in groups of hundreds and sometimes even thousands. There are many terms used to describe groups of cetaceans; a large gathering may be called a school or a herd. Orca groups, which generally remain together throughout their lifetime, are referred to as pods. The term 'pod' is also applied generically in other instances, such as 'a pod of dolphins', while female–calf groups are often referred to as nursery pods. As a general rule:

• Most dolphins live in fission–fusion societies, which consist of social groups ranging from a few animals to more than a thousand individuals. The structure of these groups is very fluid and changes frequently; sometimes there will be hundreds of animals gathered together, and at other times they will split into small, scattered groups. The composition of the social groups may change daily or even hourly.

• Mature male sperm whales generally live alone, while younger males may be sighted in pairs or small groups. Females and their young remain in larger social groups known as nursery pods. Breeding males join the nursery pods during the mating season and then depart for their more solitary existence.

• Beaked whales are among the most mysterious cetaceans and are seldom observed. Most recorded sightings are of one or two animals or small groups.

• Porpoises appear to live, for the most part, singularly or in small fluid social groups. Only a few of the seven porpoise species have been well studied, so knowledge of their social behaviour is somewhat limited.

• Most baleen whales come together on their winter breeding and summer feeding grounds but will spend the rest of their time alone, with their offspring or in small groups.

Some dolphin species have been observed in groups that range from a few to even thousands of individuals.

Members of a sperm whale nursery pod.

COURTSHIP AND MATING

Cetacean courtship ranges from the elaborate to the minimalistic. Some species engage in long 'seductions', involving touching and chasing, while others have no obvious preliminary rituals, happily mating with different partners over the course of just a few minutes. Some species, such as the rarely sighted beaked whales, pygmy right whales, dwarf sperm whales and others, are seldom observed, so their courtship and mating behaviours remain, for the most part, a mystery.

MATING AMONG ODONTOCETES

Sperm whales

In sperm whale societies, the choice of breeding partner seems to be made by the females, which also tend to opt for mature males: the males are not acceptable as breeding partners until they are in their early to mid-20s. When these mature and very large whales arrive at their tropical breeding grounds, the females have different reactions. In some cases, excited females have been observed emitting numerous coda sounds, and then rushing over to a bull and swarming around him, rubbing themselves along his huge body. Conversely, researchers have observed females completely ignoring other bulls, even diving down and swimming away from them. It appears the males display

Male and female cetaceans can obviously tell one another apart, but human researchers must be able to see the undersides of the animals they are observing in order to determine their sex, as cetacean sex organs are hidden inside a genital slit. In females, the genitals and anus are close together, with a mammary slit located on either side (right). In males, the genital slit is further forward from the anus and there are no mammaries (left).

little aggression towards each other; rather, they spend much of their time swimming from one group of females to another, often mating with a number of different females.

Dolphins

Dolphin species living in social groups have sexual encounters throughout the year, but this tends to increase when numbers of individuals come together. Courtship may involve gentle nudging and stroking but more often seems to elicit boisterous behaviours such as rapid chasing, lob-tailing and leaping. Noisy group leaps are common in many species when sexual activity is taking place.

Sexual behaviour does not, however, always lead to copulation, and young males in particular have been observed engaging in socio-sexual activities with one another. In most dolphin societies, sexual interactions become more frenetic and intense during periods when conception is likely to occur. In some bottlenose populations, for example, males actively work together to 'kidnap' certain females in order to breed with them at these times. While most cetaceans mate only with their own species, New Zealand dusky dolphins have been observed mating with common dolphins.

Killer whales

Multi-pod killer whale groups are often observed during the summer and early autumn, and it is believed that mating takes place outside the family pod during these encounters.

A single female dusky dolphin may be pursued by six or seven males, but she often mates with only one or two of them. The mating may appear to be random, but it is also possible that a female is choosing partners based on their ability to impress her with their agility and speed.

Killer whales generally live in the same family pod throughout their entire life and mating is thought to take place when pods meet in summer and early autumn.

Mating among dusky dolphins is not always a frenzied affair. At certain times it is very relaxed and slow, with the appearance of being tender.

Head-first re-entry leaps commonly occur in dusky dolphin mating groups, with the female leaping first followed by one or two males. Researchers believe the leaps may be part of the female's avoidance strategy, allowing her to catch her breath and then dive efficiently back down through the water.

MATING AMONG MYSTICETES

Many baleen species migrate annually between summer feeding grounds near the poles and winter breeding grounds in more tropical waters. Grey, humpback and right whales migrate annually to known breeding grounds. Interestingly, even with all the cetacean research that has taken place, the mating grounds of some baleen species, such as blue and fin whales, are for the most part unknown. There has even been a suggestion that some populations of blues and fins do not have specific breeding grounds but instead search out mates across vast expanses of ocean. This may be because their worldwide populations have been reduced to small numbers.

Humpback whales

Humpback whales are not always gentle giants while in their tropical breeding grounds. There may be quiet moments when males are singing or benignly swimming alongside females, but when groups of males are actively courting and attempting to mate, the singing stops and the peaceful behaviour disappears. The rambunctious action of courting humpbacks is usually accompanied by lots of noise, with the males snorting, trumpeting and producing loud, aggressive-sounding bellows. Males will attempt to mate with a number of different females, and research suggests that some males may form coalitions in order to secure a particular female and mate with her. Pregnant females are the first to leave the breeding grounds in order to have the longest possible time on their feeding grounds near the poles in preparation for their return journey back to the tropics to give birth.

Right whales

Most right whale populations mate and calve in latitudes of about 30°N or 30°S, although a few populations also utilise waters around subantarctic islands. Right whale females engage in a multiple mating system, copulating with different males one after the other. Their courtship rituals range from stroking and rolling around each other to rapid chasing. Female right whales normally give birth in winter after a gestation period of approximately 12 months, so even though they engage in sex year-round it appears that conception occurs mainly on their winter breeding ground.

FASCINATING FACT

Right whale males have the largest testes (up to 1 tonne) and the longest penis (around 2.3 metres long) of any animal.

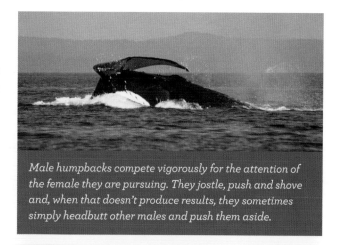

Male humpbacks compete vigorously for the attention of the female they are pursuing. They jostle, push and shove and, when that doesn't produce results, they sometimes simply headbutt other males and push them aside.

When a female right whale wishes to avoid amorous males, she turns belly up on the surface. The males will then hover around her, sometimes pushing and shoving, and as soon as the female stirs slightly to start to turn back over, they will grab a quick breath and dive under her, hoping for the chance to mate.

Some of New Zealand's southern right whales use the cold subantarctic waters around Campbell and Auckland islands as their winter breeding grounds.

BIRTH AND PARENTHOOD

The gestation periods of cetaceans vary: most dolphin calves are born after a gestation of 9–10 months; baleen whales have a gestation period of 11–12 months; and killer and sperm whales have longer gestation periods of 14–16 months. Whale and dolphin calves are born underwater and delivery normally occurs quickly, but it can also take hours. Once the calf emerges, the mother quickly assists it to the surface for its first breath of air. New calves are miniature replicas of their parents except for their floppy dorsal fins and tail flukes, which stiffen up within the first few weeks.

All female cetaceans have two milk glands hidden inside the mammary slits on their stomach, and they nurse their calves by squirting milk directly into their mouths. Hungry calves have a feathery-tipped tongue that helps create a good seal when they attach themselves for a feed. The calves grow quickly on their diet of rich, fat-laden milk – in the first few months of its life, a blue whale calf will grow almost 4 centimetres and gain about 90 kilograms a day.

Most newborn cetaceans are around a third the size of their mothers. Calves nurse for six months to over a year, although some species, such as sperm whales, will continue to suckle occasionally until they are teenagers. A calf is weaned gradually, with a period of overlap when it is nursing and feeding on solid food at the same time. In addition to learning how to feed itself, a calf must learn the social and behavioural skills it will need as an adult. Calves remain with their mothers for months and, in some cases, years, depending on the species.

Whale and dolphin calves are born tail first, perhaps so that they won't drown in the event of a long or difficult birth.

A new whale or dolphin calf will remain close to its mother's side for the first few days after birth.

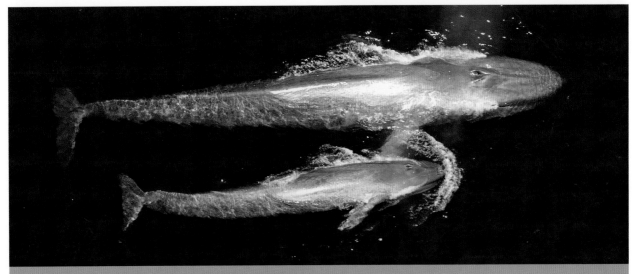

A blue whale mother gives birth to the largest baby in the world – her newborn calf may be 7 metres long and weigh up to 3 tonnes.

A humpback whale in Antarctic waters displays its tail, which is covered with a yellow film from the bodies of thousands of diatoms, single-celled phytoplankton, which proliferate during the long daylight hours of summer.

MIGRATION

All cetaceans must find suitable locations for feeding, mating and calving, which means that they are almost always on the move. Baleen whale species such as grey, blue, fin, minke and humpback migrate many thousands of kilometres between their high-latitude feeding grounds in the Arctic or Antarctic and their low-latitude mating grounds in tropical waters. Some Bryde's whales also migrate, while other populations remain in tropical or semi-tropical waters year-round. And while most minke whales migrate to lower latitudes for breeding, some are known to overwinter in Antarctic waters. Other mysticetes make shorter migrations or simply move around in localised areas.

Southern hemisphere humpbacks feed during summer on the vast quantities of krill that appear in Antarctic waters. As autumn approaches, the whales begin their long journeys to warmer waters, where they will mate and calve. Humpbacks swimming north through New Zealand's

Some like it cold! Northern hemisphere bowheads remain in Arctic waters year-round, while belugas and narwhals live in both Arctic and subarctic waters. Some killer whale populations live in the far northern waters of the northern hemisphere, while others live year-round in Antarctic waters, merely moving back and forth as the ice dictates.

waters will spend their winter months around South Pacific islands, such as Tonga and New Caledonia.

Right whales normally mate and calve in latitudes of around 30°N and 30°S, and then migrate to higher latitudes around 50–60° to feed. A few populations, such as New Zealand's southern right whales, mate and calve around subantarctic islands.

Most toothed cetaceans have home ranges and undertake somewhat limited journeys. For example, New Zealand's killer whale population travels around both the North and South islands in the quest for food, while most Hector's dolphins remain in very localised areas, rarely travelling further than 30–60 kilometres from their home base. Many dolphin species live within a relatively definitive range; when and how far they travel is mainly dependent on the availability of their prey. While female sperm whales do not travel great distances, many males make far longer journeys. Large bulls are known to travel to cold polar waters in search of food, although in some areas close to the Equator there is no clear seasonal migration.

Migrating whales need to know when to leave and where to go. How they accomplish this remains, for the most part, a mystery. In birds and mammals, the pineal gland produces more melatonin in darkness and less in daylight hours. In whales, the gland may act as a kind of timekeeper, giving them cues as to when to leave a feeding or breeding ground. For example, when daylight hours decrease as winter approaches in polar regions, more melatonin is produced, which might signal that it's time to depart for the breeding grounds.

Researchers have noted migrating humpback and grey whale individuals in the same locations from one year to the next. The navigational methods used by whales are still unclear but it is suggested that those migrating close to land may follow the contours of the coastline, while other theories suggest that whales use celestial navigation, ocean currents or the Earth's magnetic field to guide their way. It is intriguing that monitored humpbacks travelling from Hawai'i to southeast Alaska follow a track that is within 1° of magnetic north.

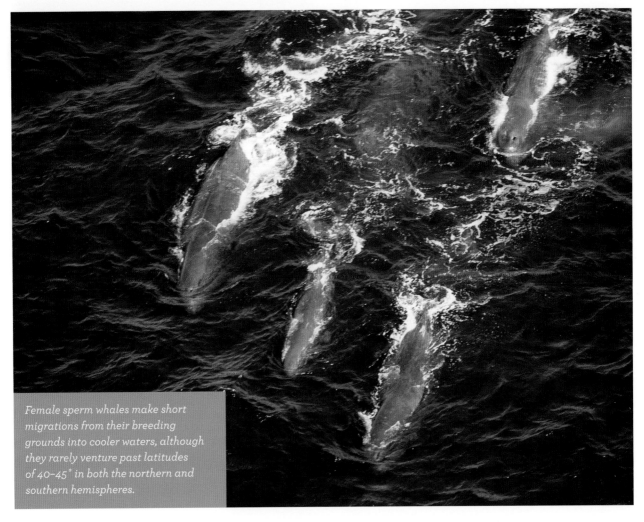

Female sperm whales make short migrations from their breeding grounds into cooler waters, although they rarely venture past latitudes of 40–45° in both the northern and southern hemispheres.

SURFACE BEHAVIOURS

Cetaceans exhibit a number of different behaviours at the surface of the sea. Although these are easily observed and have been well documented by researchers, their purpose is not always understood.

BREACHING

When cetaceans launch themselves head first into the air, they are said to be breaching. At times, a breaching whale may completely clear the water with its body, while at others, only a proportion of its body appears. Porpoises are seldom observed breaching while many of the large whales such as blue, fin, sei and sperm breach only occasionally. Humpbacks and most dolphin species are often observed leaping from the sea and when one animal starts to breach, others will often follow suit.

Breaching is thought to have a number of purposes: it may help the whales to dislodge barnacles or other parasites; it may be a form of communication, such as signalling a whale's presence; it may be a form of intimidation, indicating that the whale is powerful; or it may assist feeding cetaceans by herding fish or other schooling prey closer together. There are times, however, when it appears the whales are breaching simply for fun.

Some dolphin species are particularly acrobatic. New Zealand's dusky dolphins fit into this category and have a large repertoire of leaps, including full somersaults that may be repeated again and again.

LOB-TAILING AND FLIPPER-SLAPPING

Like breaching, these active and noisy behaviours occur frequently in both feeding and mating grounds. Lob-tailing whales repeatedly bang their tails on the water surface while keeping their heads and blowholes submerged, while flipper-slapping whales lie on their sides and smack the water with their flippers. Both behaviours create a loud cracking sound, which is proposed to advertise the whale's location, act as a signal of aggression, intimidation or displeasure, or assist in herding fish during feeding.

SPYHOPPING

When a cetacean wants to observe the world above water, it lifts its head out vertically and spyhops. Cetacean eyes can adapt to see both in air and in water, and studies on captive dolphins have shown that their vision out of water is well developed. It is not believed that cetaceans see colours as we know them, but they may be able to distinguish some shades in the blue–green range of the colour spectrum.

A whale will sometimes lob-tail continuously for long periods of time. This New Zealand southern right whale managed more than 50 lob-tails in a one-hour period while in its winter breeding ground.

When cetaceans are in a hurry, like these dusky dolphins, they literally leap out of the water repeatedly, 'porpoising' parallel to the surface each time they take a breath. The whalers used to say that porpoising dolphins were running.

This curious humpback calf spyhopped three times in order to check out the strange humans who had entered its world.

BOW- AND WAKE-RIDING

Bow-riding is a behaviour where dolphins, porpoises and occasionally other smaller toothed whales hitch a free ride in the pressure wave created at the front of a boat as it travels through the water. Sometimes these smaller cetaceans will even ride the bow wave of large whales as they travel through the sea. Cetaceans may bow-ride to conserve energy, but it is believed that they mostly do it as a form of play – they are often observed racing over to a boat for a ride and then returning to the spot from where they came.

Dusky dolphins bow-riding in front of a tourism vessel near Kaikoura.

Sometimes cetaceans will ride behind a boat in its frothy wake. Wake-riding is usually a rambunctious pastime, during which the animal will surf, twist and turn in the waves created by the boat's engine, and will even partake in a kind of whirlpool bath by turning upside down and swimming in the bubbles.

WHALE
STRANDINGS

My breath was taken away by the animal laid out before me:
so powerful, yet so vulnerable and exposed.

Renee Kelly, Project Jonah volunteer (2008)

When a single whale strands, it is often the result of illness, injury or death. Some causes are natural: a whale might be infected with parasites or disease, it might have had problems with birthing, or it might have been attacked by killer whales or sharks. Young calves might have lost their mother, while older whales might simply have died of old age. Unnatural deaths also occur, such as entanglement in fishing gear, boat strikes and pollution – a number of stranded whales and dolphins have been found with ropes, plastic bags and other ocean debris in their stomachs.

A cow–calf pair that strands together is considered a single stranding; otherwise, any stranding involving two or more individuals is considered a mass stranding. The exact cause of mass strandings is still the subject of much conjecture. One common feature is that they often occur among highly social, toothed whale species, such as sperm and pilot whales, which use echolocation to navigate in deep, open ocean waters. The following factors are believed to contribute to mass strandings:

- **Social cohesion** If one or more sick or injured member of a large pod strands, their distress calls will cause other members of the pod to come to their aid. The entire pod may end up stranding in their effort to assist their helpless companions.

- **Atmospheric conditions** Some scientists believe that whales use Earth's magnetic field as a map when they are navigating. There is evidence to suggest that extreme weather such as electrical storms or events such as sunspots, which affect Earth's magnetic field, may lead to navigational errors that cause whales to swim too close to land.

- **Underwater explosions** These can rupture the eardrums of whales, which may in turn affect their ability to communicate, hunt or navigate. Mass strandings have been known to occur after underwater seismic tests.

- **Physical characteristics of the land** Coastal areas with sloping, sandy ocean floor bottoms, combined with sand spits or a long finger of land jutting into the sea, appear to act as natural whale traps. When the whales end up inside the long finger of land, the sloping, sandy ocean floor absorbs their sonar waves, thus reducing the effectiveness of their echolocation system, and they become confused. Adding to this problem is the fact that these areas usually have extreme tidal fluctuations; many strandings occur during periods of maximum high and low tides.

Every year, a variety of cetacean species, including rare beaked whales, come ashore on beaches around the country. Most incidents are of single animals but mass strandings also occur, sometimes involving hundreds of individuals. New Zealand has one of the highest stranding rates in the world. From August 2010 to March 2011 alone, there were 125 recorded stranding events involving more than 500 individual whales or dolphins.

The coastline of New Zealand is long and convoluted. The country's land masses spread from the northern subtropical Kermadec Islands to the subantarctic islands thousands of kilometres further south. Deep water comes close to shore in many areas, while in other places, beaches with tidal extremes challenge whales by providing swimmable water at high tide and a sandy beach on which to strand at low tide. Along the coast are a number of stranding hotspots – areas where whales and dolphins consistently beach themselves.

BARBARA TODD:

Diary of a whale stranding

Imagine walking down to a beautiful sandy beach... and finding the scattered bodies of more than 340 stranded long-finned pilot whales lying in the shallow waters. This is the sight that greets me and my fellow rescuers one January morning in 1991 in Golden Bay, at the top of New Zealand's South Island. The whales seem to be everywhere and the initial feeling is one of helplessness. A few whales have already died; others have rolled onto their sides and are struggling in vain to right themselves again. A tiny calf calls for its mother, which lies 20 metres away, thrashing her tail in frustration as she attempts to reach her young baby.

DAY 1 11:00 AM

The tide is low — there are almost six hours to go before high tide, when the first attempt can be made to refloat the beach-cast whales. It's time for the rescuers to go to work! Black whale bodies soon absorb the sun's heat, so our first job is to cover the whales with sheets and pour cool water over them. A few volunteers must be reminded not to pour water down the whales' blowholes. New rescuers continue to arrive. Each person is assigned a whale or a small group of whales to look after. The rescuers' job is to talk quietly to their whales, keep them cool, and help to upright them before the tide moves in; a whale that is trapped on its side will drown once its blowhole is covered with water.

DAY 1 4:30 PM

At last high tide approaches. By now, every whale has at least one human minder and everyone works to get their whale ready to swim to safety. As the whales are refloated, the rescuers rock them gently from side to side to help them regain their equilibrium. Finally, it's the moment that everyone has been waiting for. The minders walk their whales into deeper water and set them free. Some whales swim away, but others, confused and tired, return to the beach. The restranded whales call out in distress, and within an hour all the whales that swam to safety have returned to their mates.

DAY 1 7:30 PM

Once again the entire pod lies helpless
on the golden sand. Frustration and a
sense of failure take hold as the tide
drops and night approaches; it will be
10 long hours before another rescue
attempt can be made. All the volunteers
are weary, and we grab a hot drink before
returning to talk quietly to our whales.

DAY 2 6:00 AM

After a seemingly endless night, it is
dawn and another high tide approaches.
There are tears; more whales have died
and both people and whales are exhausted.
But then, a surge of adrenalin seems to
flow through everyone — this tide is a
very high one and a feeling of hope is
in the air. The whales are refloated and
grouped together. Slowly the rescuers

wade into deeper water, gently rocking
their whales as they guide them towards
the open sea. The water is now almost
around the lead rescuers' necks, and it's
time to set the whales free once again.
A final hug, a last stroke and then everyone
holds their breath. Some whales, confused,
start to swim back to land, and people
dive into the water to turn them around.

Suddenly, the whales seem to find their
way and start moving toward the horizon.
The mother and calf have been reunited
and are swimming strongly with the rest
of the pod. Cheers and tears flow
simultaneously as rescuers bid the
whales farewell.

This stranding ultimately resulted in one
of New Zealand's — and the world's — most
successful whale rescues. Of the 340-plus
whales that stranded, 294 were saved and
swam to freedom.

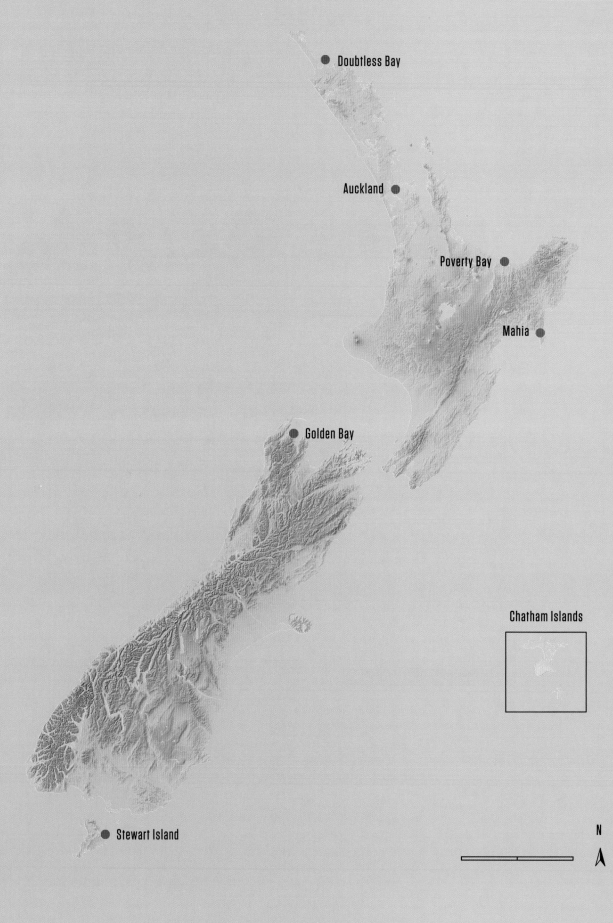

Doubtless Bay

Auckland

Poverty Bay

Mahia

Golden Bay

Chatham Islands

Stewart Island

N

AOTEAROA NEW ZEALAND STRANDING HOTSPOTS

The map opposite indicates the stranding hotspots around the country.

NORTH ISLAND

Karikari, Doubtless Bay

The shallow, sandy beach of Karikari has been the site of frequent mass strandings of long-finned pilot whales.

Auckland's west coast

This area has often been the location of single and mass strandings of sperm whales.

Wainui, Poverty Bay

One of New Zealand's largest strandings of sperm whales occurred at Okitu Beach in Poverty Bay in March 1970. All the whales subsequently died; their marked graves can still be seen.

Mahia, Hawke's Bay

The deep-water trenches off Mahia Peninsula are the main home of New Zealand's pygmy sperm whales. At least 80 per cent of New Zealand's pygmy sperm whale strandings occur on Mahia's beaches.

SOUTH ISLAND

Golden Bay

The bay is a classic example of a whale trap. Farewell Spit juts out at one end, blocking access to the open sea, and the rest of the bay is long and curving, with shallow, sloping beaches and extreme high and low tides. Although Golden Bay is renowned for mass strandings of long-finned pilot whales, it has also been the stranding site for at least 14 other whale and dolphin species.

Stewart Island/Rakiura

With its numerous shallow, sandy beaches, Stewart Island/Rakiura has been a hotspot for strandings of migratory and cold-water whale species.

Chatham Islands

Lying almost 700 kilometres east of the South Island, the Chatham Islands are located in an area of converging currents that produce abundant food, this in turn attracting a large diversity of marine species. Additionally, the islands have a number of shallow, sloping beaches adjacent to deep water. These combined factors have led to a high incidence of whale strandings of many different species. The first stranding recorded in the New Zealand database occurred on the Chatham Islands in 1840 and one of New Zealand's largest recorded strandings occurred there in 1918, when over 1000 long-finned pilot whales stranded on Long Beach on Chatham Island (Rekohu).

REACTIONS TO STRANDINGS IN AOTEAROA NEW ZEALAND

For early Māori settlers meat from stranded whales provided much-needed protein in a country devoid of large land mammals. Whale flesh was hung to dry or cooked in a hangi (earth oven). Whale oil was used for polish or scent, while whale bone and teeth were used to make weapons, fishing hooks, musical instruments and ornaments (see pages 102–7).

The arrival and establishment of European whalers altered Māori perspective to stranded whales. While meat and bones were still utilised, Māori also partook in small-scale commercial activities, boiling down the blubber of stranded whales in order to sell it, and joining up with whalers or establishing their own whaling operations to hunt for whales.

By the 1950s, responses towards stranded whales had shifted, and they were basically regarded as a public health problem to be disposed of as quickly as possible. New Zealand's last whaling station closed in 1964, and not long after, people began to look at whales in a new light – as creatures that should be protected, as opposed to creatures that offered commercial gain.

During the 1970s, 'Save the whales' became a rallying cry for conservationist organisations such as Greenpeace and the World

Māori whalers boiling down blubber, around 1900.

The maihi (bargeboards) of many pātaka (storehouses, above) often had intricate pakake, or whale-like patterns, carved in them. Pakake carvings consist of a large scroll-like shape that represents a whale's head, with a body shape tapering down toward the tail carved beneath the head. A pātaka stored with whale meat could provide a tribe with food for many months.

Wildlife Fund. Project Jonah, a New Zealand anti-whaling organisation founded in 1974, added another string to its bow in 1976, when it embarked on a campaign to rescue stranded whales and return them to the sea. In 1985, Project Jonah designed and developed an inflatable pontoon system, the world's first whale rescue flotation device. Since the mid-1980s, rescue techniques have been greatly refined, and Project Jonah has trained large numbers of 'PJ medics', who have assisted in the rescue of many hundreds of stranded whales.

The current attitude towards stranded cetaceans is to help them as much as possible. Communities often work together to save or aid whales that beach on New Zealand shores. When a stranding occurs, the Department of Conservation (DOC), which is responsible for all New Zealand marine mammals, works alongside local iwi (tribes) to ensure that Māori cultural beliefs and needs are respected. Project Jonah volunteers and members of the local community often assist in the rescue efforts.

While strandings are distressing events for many observers, they also present a window of opportunity for scientists to study the biology of the whale species and learn new information. Museums also benefit from stranded whales. The Museum of New Zealand Te Papa Tongarewa in Wellington, for example, has one of the largest collections of beaked whale skeletons in the world. For some species, such as beaked whales, virtually everything that is known about them comes from studies of stranded specimens.

WHALE STRANDING PIONEER

One of the first New Zealanders to become actively involved in whale strandings was Napier resident and commercial fisherman Frank Robson. Robson did not buy into the popular theory that stranded whales were committing mass suicide, and during the mid- to late 1950s, he not only looked at possible causes for whale strandings but also developed techniques to assess and care for the animals while they were on land as well as ways of returning them to the sea. Between 1958 and 1982, Robson recorded information from 24 mass strandings involving more than 1500 individuals as well as an additional 132 stranded individuals. His pioneering work laid the groundwork for the methods used on stranded cetaceans by the conservation groups and researchers who were to follow. In his 1984 publication *Strandings: Ways to Save Whales*, Robson explained his dedication: 'Just now, when the numbers of whales and dolphins are lower world-wide than at any other period of history due to deliberate and incidental killing by men, the use of every opportunity to save some of them from death by stranding is some small atonement.'

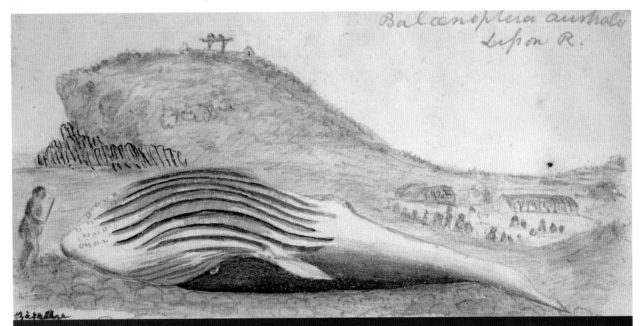

People gather in 1841 to feast on this whale that has just come ashore. To Māori, stranded whales were seen as a koha (gift) from Tangaroa, the god of the sea.

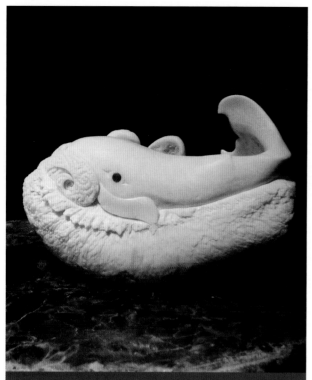

For artisans like Brian Flintoff, carving a whale bone enables him 'to take part of a whale and turn it back into a whale'. When Brian was gifted two ear bones from a stranded humpback whale, he left one in its natural state and carved a whale from the other, using the bone's natural contours in the carving.

In 1985, Project Jonah developed an innovative pontoon system that could be used when rescuing stranded whales or dolphins. The pontoons are used to tow animals out to sea; sometimes single animals are taken out by this means, at other times an individual from a mass stranding may be towed out to sea in the hope that others will follow.

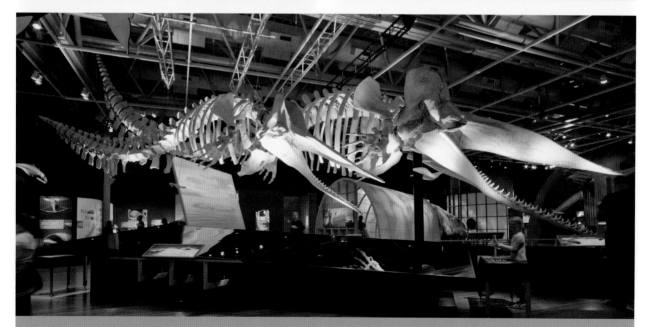

In 2003, 12 sperm whales stranded at Karekare beach off Auckland's west coast. To Māori the stranding was seen as a tohu or sign. The largest of the males was gifted to the national museum, Te Papa, by the local Māori tribe, Te Kawerau-a-Maki. The whale was given the name Tū Hononga, and his skeleton (right) now hangs at Te Papa.

ASSISTING IN A WHALE STRANDING

Whatever the reason for a stranding, certain steps can be taken to assist the whales as they lie helpless on the beach. The first step is to make sure that DOC is aware of the stranding so that its staff can assemble the best team possible to ensure the safety of both whales and volunteers. DOC needs to know the exact location of the stranded whales, the number and species (if known), and the sea and tidal conditions. All decisions regarding the rescue, and whether or not euthanasia is required, will be made by DOC rangers in consultation with local iwi. Volunteers need to be aware that they will probably be spending long hours in the elements. Warm clothing, sunblock and drinking water are essentials, as are rest periods to ensure they do not become overtired themselves. There are then some basic dos and don'ts that rescuers should follow when working with stranded animals:

DON'TS

1. Don't cover or pour water or sand down the blowhole, and don't cover a whale's eyes.

2. Don't step over or straddle the tail or pull on the flippers, dorsal fin or tail flukes.

3. Don't drag or roll the whale to the water, as the flippers and dorsal fins are easily dislocated.

DOS

1. Do keep stranded whales cool. If the whales are high and dry, the first priority is to keep them from overheating by pouring water over them. A whale's black skin absorbs heat from the sun; covering its body with wet sheets will also help it stay cool and protect its delicate skin from sunburn.

2. Do keep the whales as comfortable as possible. If whales are lying on their side, they need to be gently moved into an upright position. This can be achieved by digging a shallow trench alongside the whale and then rolling the animal into it. Sand or sandbags can then be used to help prop the whale upright and shallow indentations can be dug under the flippers so they can hang more freely.

3. Do stay calm. The whales will be badly stressed, so it is important to talk quietly and move slowly so as not to frighten them further. The whales will be less stressed if they are able to see the people who are trying to help them, so it's best to avoid standing or sitting directly in front of the head.

4. Do rock the whales when they are refloated. Whales that lie stationary on land soon lose their equilibrium, and when the tide comes in and they are refloated, they often have a hard time staying upright and will flop over onto their side. Gently rocking the whales from side to side will help them to restore their sense of balance.

A whale stranding often brings people together in a common effort to assist their fellow mammals. For these rescuers, it can be a very emotional experience. Most people bond with a particular whale or whales during the rescue attempt. If the animals are saved, there is great joy; conversely, there is great sadness when whales die either naturally or when euthanasia is required.

ANTON VAN HELDEN:
Collecting stranded whales

I find whales pretty remarkable. Even though they are among the largest animals on this planet, we still know very little about most of them; how they live, what they do and where they come from. In 1989, I was employed by the National Museum, where I worked with cetaceans until February 2013.

New Zealand's national museum, now known as Te Papa, has been collecting specimens from stranded dead whales since 1865 and currently holds one of the world's largest marine mammal collections. This includes the largest collection of beaked whales in the world, comprising skeletons and other preserved tissues. Three beaked whale skulls are pictured above. For more than 20 years, I also managed the New Zealand Whale Stranding database for the Department of Conservation (DOC). Information gathered at stranding events helps to define the different species that exist in New Zealand's waters, and during my time at Te Papa I helped in the recognition of nine cetacean species that we now know belong to the biota of New Zealand. Some species are seldom or even never observed in the wild. A stranding can open a rare window into the lives of these animals; it is our only opportunity to really get inside these whales and have a look at how they function. For me, that is truly exciting.

As part of my role at Te Papa, I collected skeletons and tissues from stranded animals in order to help understand their species. The difficulty with collecting whales is that they are large animals, — if they are too big, they can't be brought back to a lab. Quite often I would find myself — along with other Te Papa staff, DOC colleagues and local iwi — on a wild stretch of coast, dressed from head to foot in protective gear to minimise the risk of infection from the bacteria and muck we frequently had to work in. It was often windswept and raining or hailing, or under scorching sun, with the tide rapidly coming in as we raced against time trying to collect vital bits of information, for even the smallest scraps can provide new insights into the lives of these animals. Whenever possible, we would dissect freshly dead animals and record information about them using photographs, drawings and notes on our observations. Tū Hononga, a stranded sperm whale, is pictured opposite.

I focused primarily on collecting rare and unusual species that strand on the New Zealand coastline, species poorly represented in other worldwide collections, and I have been called on by people all over the world to help identify whales. In 2002, I was involved in the discovery of a new species, Perrin's beaked whale, and in the same year I also led a team that resurrected the spade-toothed whale, a species known only from a handful of specimens worldwide. I also helped to describe the colour patterns of both the spade-toothed whale and Shepherd's beaked whale.

Another project that spanned a number of years involved collecting specimens of the pygmy right whale. Collaborative research on this species has revealed remarkable insights into its biology,

functional anatomy and evolutionary origins. The recent discoveries illustrate the importance of museums in keeping specimens in perpetuity, so that as new technologies develop researchers can use them to discover fresh information about the different species represented in the collection.

Working collaboratively with iwi has always been important to me. I have sought to ensure their mana taonga with collected specimens and helped to forge strong relationships between iwi and the Te Papa collection. I remain greatly indebted to these communities for the tremendous belief they have put in Te Papa to care for their taonga and for allowing their stories to be told.

One of my proudest achievements while at Te Papa was to instigate and develop, as subject expert, the exhibition Whales | Tohorā, which was first shown in 2007 and has gone on to tour North America, attracting more than a million visitors. When I was a child I wanted to see large whale skeletons on display at the museum and they did not have them, so a personal triumph for me was to get the pygmy blue whale on display at Te Papa and the two touring sperm whale skeletons in the Whales/Tohorā exhibition (Tū Hononga and Hine Wainui).

Although I am no longer at Te Papa, my work with whales is not over. These magnificent creatures deserve our attention and I will continue to encourage research on and appreciation of them.

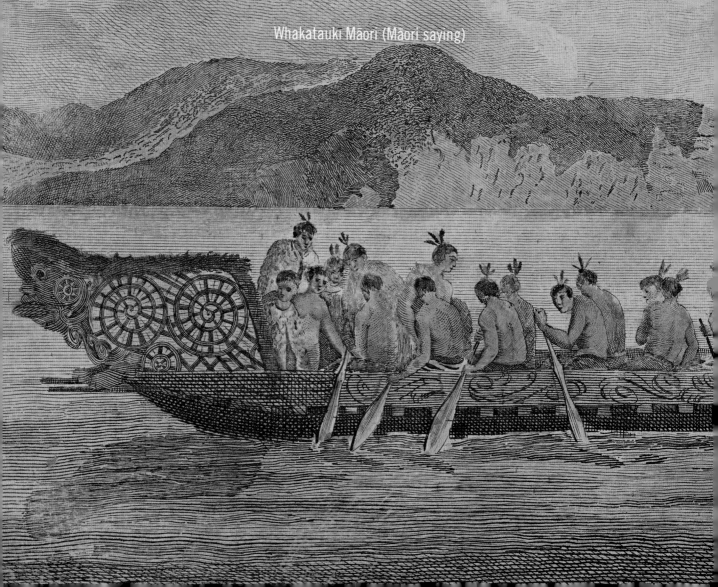

PART TWO

OF WHALES
AND MAN

Tere tohorā, tere tangata.
Where whales journey, people follow.

Whakataukī Māori (Māori saying)

WHALES
AND PEOPLE

Nature is a kind of art sans art;
and the right human attitude to it ought to be,
unashamedly, poetic rather than scientific.

John Fowles, *The Blinded Eye* (1984)

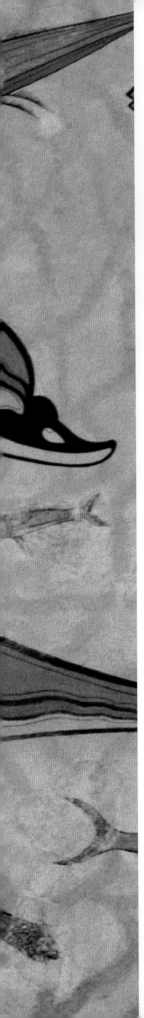

EARLIEST CONNECTIONS

Throughout the ages, whales and dolphins have been a focal point of social, cultural and religious life in coastal societies around the world, and they are present in stories and artwork dating back to at least 1500 BC.

Some of the oldest known accounts and depictions of cetaceans come from the Mediterranean region, where dolphins appear in numerous myths. The Greek god Poseidon was led to his true love by a dolphin, and to show his gratitude he formed the constellation Delphinus. In another legend, Poseidon saw his son Taras drowning and sent a dolphin to save him. The town of Taras, founded in 700 BC, is located in the place where the dolphin is supposed to have deposited the boy, and for generations a dolphin rider was the predominant image on its coins.

Ancient Greeks and Romans both believed that dolphins were created to be the friends and helpers of mankind, and that killing a dolphin was as much a crime as killing a person. There are many stories of dolphins befriending humans, and the artists of the time, inspired by accounts of the intelligence and 'kindness' of dolphins, incorporated the animals into their drawings, paintings, mosaics and sculptures. Images of dolphins also featured on many coins, which were thought to provide safety to travellers.

The Greek philosopher and scholar Aristotle (384–322 BC) was one of the first writers to recognise the dolphin's mammalian characteristics, even though he continued to refer to the animals as 'fish'. Roman author and scientist Pliny the Elder (AD 23–79) often wrote of whales and dolphins, and seemed particularly fascinated with stories recounting relationships between dolphins and humans. In one account, he describes fishermen working together with dolphins to catch fish, an occurrence recorded in many different times and cultures around the world.

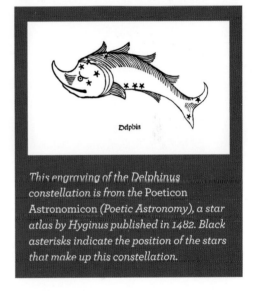

This engraving of the Delphinus constellation is from the Poeticon Astronomicon (Poetic Astronomy), a star atlas by Hyginus published in 1482. Black asterisks indicate the position of the stars that make up this constellation.

FASCINATING FACT

The root of *delphis*, the Greek word for 'dolphin', is *delphys*, which translates as 'womb'. The original meaning for *delphis* could therefore have been 'fish with a womb'.

There are also tales of early Christians who were initially saved from martyrdom by dolphins. Callistratus, a soldier in Carthage in the fourth century AD, was sewn in a leather sack and thrown into the sea when his commanding officer discovered he was a Christian. The sack was torn open by a sharp rock and Callistratus was carried ashore by dolphins, inspiring 49 other soldiers to convert to Christianity. Basil the Younger, also persecuted for his Christian beliefs and thrown into the sea in the tenth century AD, was similarly rescued by dolphins and ultimately carried to shore near Constantinople. Both men later died for their beliefs and became saints.

Relationships between humans and whales are less common in stories, perhaps due to the sheer size of whales, which makes them more intimidating. The biblical description of Leviathan in Job (41:25) portrays a somewhat terrifying creature – 'when he raises himself up the mighty are afraid; at the crashing they are beside themselves' – and in fact many of the early stories of whales refer to them as 'sea monsters'. In a well-known biblical story, Jonah is swallowed by a whale for disobeying God's command, but when he prays for forgiveness the whale finally spits him out after three days.

In his history of Alexander the Great's 327–326 BC campaign to India, *Indica*, Greek historian Arrian of Nicomedia (*c*. AD 86–160) recorded an incident involving whales in the Indian Ocean:

> Nearchus states that when they left Cyiza, about daybreak they saw water being blown upwards from the sea as it might be shot upwards by the force of a waterspout. They were astonished, and asked the pilots of the convoy what it might be and how it was caused; they replied that these whales as they rove about the ocean spout up the water to a great height; the sailors, however, were so horrified that the oars fell from their hands.

This fear was not relieved by scholars such as Pliny the Elder, who in *Naturalis Historia* (*c*. AD 77–79) described whales as 'a mighty mass and lump of flesh without all fashion, armed with most terrible, sharpe, and cutting teeth'. He also wrote, 'The Indian Sea breeds the most and biggest fishes that are…the Whales and Whirlpools called Balaena take up as much in length as four acres or arpens of land.'

Other stories also relate to the enormous size of whales, which are sometimes mistaken for islands on which unwitting sailors are drawn ashore, only to discover that their safe haven is actually the back of a living sea monster that dives beneath them, sinking all of their ships. In the sixth century AD, Saint Brendan set sail from Ireland

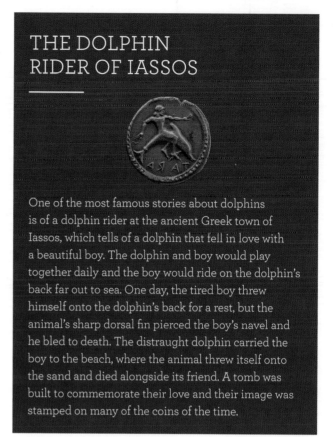

THE DOLPHIN RIDER OF IASSOS

One of the most famous stories about dolphins is of a dolphin rider at the ancient Greek town of Iassos, which tells of a dolphin that fell in love with a beautiful boy. The dolphin and boy would play together daily and the boy would ride on the dolphin's back far out to sea. One day, the tired boy threw himself onto the dolphin's back for a rest, but the animal's sharp dorsal fin pierced the boy's navel and he bled to death. The distraught dolphin carried the boy to the beach, where the animal threw itself onto the sand and died alongside its friend. A tomb was built to commemorate their love and their image was stamped on many of the coins of the time.

with other monks to find the Promised Land. During their journey, they came upon a stony barren island, but when they lit a fire to cook their meat the 'island' began to move, causing the terrified monks to rush back to their waiting boats. In this story, the monks escape, and later the whale occasionally visits them; one year, they even celebrate Easter mass on its enormous back.

Of all the whale stories, perhaps the most widely read begins with the words 'Call me Ishmael'. *Moby-Dick*, based on Herman Melville's personal experience aboard the sperm whaler *Acushnet*, which hunted in the Pacific in the 1840s, was published in 1851 to poor critical response but is now recognised as a classic, and is one of the most widely read books on whales ever written. Melville's tale combines adventure and mortality alongside morality, and in between gives some astute observations on the whaling industry of the time and various whale and dolphin species themselves. Mixed with all is a form of pathos, as when Melville asks, 'The moot point is, whether Leviathan can long endure so wide a chase, so remorseless a havoc', and answers with 'Wherefore for all these things, we account the whale immortal in his species, however perishable in his individuality … If ever the world is to be again flooded, the eternal whale will still survive and rearing upon the topmost crest of the equatorial flood, spout his frothed defiance to the skies'.

Is: S. Brandano.

Cabo Finis terræ:

Hispani

Ga

Babaria

M. Canaria.

Insulæ Fortunatæ:

Cabo de No:

M. Attlas.

Africa.

This 1621 engraving by Wolfgang Kilian shows Saint Brendan celebrating Easter mass on the back of a whale.

SCRIMSHAW

There were many idle moments on board whaling ships, and some of the men occupied their time with creating scrimshaw artworks out of ivory teeth and other pieces of whale bone. Although the exact origin of the word 'scrimshaw' is unknown, the art itself is believed to have been learned from Inuit whale hunters who traded their etched pieces of whale and walrus bone with European whalers in the 16th and 17th centuries. On board a whaling ship, the men would often use a simple jack knife to scratch out a pattern, although some brought along small tools for the purpose that resembled dental implements. The whalers would first clean and polish the tooth or bone to be used and then etch out their drawings. The engraved lines would be filled in with lamp black or other colouring material and burnished with wood ash or a polish. The men also created small boxes, such as snuffboxes, from jawbones, along with knitting needles, walking sticks, buttons, napkin rings, brooches and other items. Most scrimshaw pieces were created between the 1820s and 1860s, although some work is still carried out today in cultures that continue to hunt whales or that have access to whale bone.

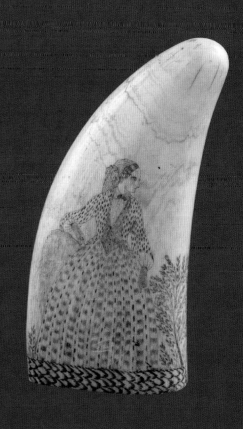

BEYOND EUROPE

In the Arctic, Inuit tribes from Alaska, Canada and Greenland have relied on whales from time immemorial to provide them with meat, blubber and oil. Whale bone was sometimes used in the construction of homes; a prehistoric qargi, or men's ceremonial house, excavated in Alaska was found to be made almost entirely from the bones of whales. The Inuit revered and hunted bowheads in particular, and often refer to themselves as the People of the Whales. Inuit life and social interactions were organised around the whale hunt, during which certain protocols had to be followed. A man's social status within his tribe was often also determined by his success as a hunter, and many Inuit believed that a whale gave itself willingly to a hunter who was worthy of its sacrifice. After a whale was killed, the Inuit of the Bering Strait region believed its spirit stayed in the body for four days, and during that time, nothing that could injure or frighten the spirit – such as pointed instruments and loud noises – was allowed in the village.

Further south, First Nations tribes living along the Pacific Northwest coast of the USA and Canada have always had strong relationships with grey, humpback and killer whales, all of which feature in their legends and artworks. Some First Nations tribes traditionally hunted migrating grey

A whale sculpture on the walls of Tan temple, a whale-worshiping temple on Ly Son Island, Vietnam. At this ornate temple near Ly Son Island's shoreline, Vietnamese fishermen offer up their prayers to an unusual god – 'Ca Ong' or Mr Whale.

whales, while other tribes fished for salmon alongside killer whales. The Haida, Tlingit and Kwakiutl cultures, to name a few, have many traditional stories, rituals and taboos surrounding whales. For some tribes, whales might change shape and become people, and people could also shape-shift and become whales. The Haida believed killer whales had the power to give them good health, strength and wealth, while the Kwakiutl would place offerings to whales and mourn the death of a whale by chanting over its body.

On the other side of the Pacific Ocean in Vietnam, fishermen believe whales and dolphins are sent by the god of the waters to protect sailors and carry shipwrecked mariners to shore. A whale or dolphin that is accidently killed in a fishermen's net is given an elaborate funeral. Three years after its death, its bones are dug up, cleaned and reassembled into a skeleton, which is then buried in a special temple. Whale temples still exist in Vietnam today; they are said to harbour the 'mandarins of the sea', who have ascended to heaven to become 'sea angels'.

In Australia, the Noonuccal tribe of Minjerribah (Stradbroke Island) had a system of sounds they used to communicate with dolphins, which would then drive shoals of fish into shallow water for the people to gather. The Wurundjeri people also communicated with dolphins using a special sequence of whistles followed by a silence, where the people and dolphins would speak 'mind to mind'. Like the Greeks, some Aboriginal tribes believe that the dolphin is the animal nearest to humankind and that to harm one would incite the anger of the spirits or the gods.

This Kwakiutl ceremonial mask depicts Kumugwe, or Chief of the Sea, an important figure in Kwakiutl mythology, the strongest power in the sea and the embodiment of great wealth. The Kwakiutl are one of some 20 diverse communities, known collectively as the Kwakwaka'wakw, united by a common language and historically centred on the coast of British Colombia.

SOUTH PACIFIC CONNECTIONS

The earliest ancestors of the Polynesians are believed to have originated in mainland Asia. One of Earth's last and greatest human migrations began around 1500 BC, when these people set off on an eastward journey that culminated in the discovery and colonisation of the islands of the South Pacific by around AD 700. This remarkable feat was accomplished without the aid of compasses, sextants, chronometers or charts.

Whales and dolphins have played an integral role in the lives of these people, whose island homes lie scattered across 40 million square kilometres of ocean. These waters host a variety of cetacean species, including pantropical spotted, spinner, striped and rough-toothed dolphins, and melon-headed, pygmy killer and short-finned pilot whales. Female sperm whales gather in deep offshore waters around islands such as Fiji and Tonga, where some give birth and others await the arrival of breeding bulls. Migratory species, such as humpback whales, travel north to Oceania's warm sheltered waters in winter in order to mate and give birth.

Traditionally, some Pacific Island communities actively hunted small cetacean species, while others opportunistically utilised stranded whales or dolphins as an alternative food source. Whale and dolphin bone and ivory also had great value. They were carved into weapons, fishhooks, musical instruments and ornaments, and were sometimes traded for other goods. At times, decorative bone or ivory objects were made specifically to commemorate a special event or ceremony.

In many of the islands, such as the Marquesas, dolphins were deliberately herded into the shore and captured. The meat from these round-ups was either eaten fresh or dried for future use, while the teeth and bones were used to make ornaments and tools. The log from a European sailing ship visiting the Marquesas in the early 1800s described one such dolphin round-up, referring to the dolphins both as 'porpoises', the common name by which they were known at the time, and as 'fish'.

> The manner of catching porpoises is truly surprising. When a shoal comes in, they get outside of them with their canoes and form a semi-circle. Then by splashing with their paddles, hallooing and jumping overboard to alarm the fish, they push for the shallow water and thence the beach, where the onshore natives pursue and take them. In this manner whole shoals are caught.
>
> David Porter, *Journal of a Cruise Made to the Pacific Ocean* (1818)

The Legend of the Voyage to New Zealand, *by Kennett Watkins (1912).*

Polynesia means 'many islands' – more than 1000 volcanic islands lie within the Polynesian Triangle.

Whale teeth were highly valued not only by the Marquesans but throughout the South Pacific. Sisi and wāseisei whale-teeth necklaces from Fiji were worn only by chiefs or other men of influence, while tabua, single ceremonial whale teeth, have long had supreme spiritual value in Fiji and are known as kava-katuranga, or chiefly items. Tabua were given as gifts at weddings, births or funerals, or as a token of sincerity in an apology. They were also used as a seal to bond a relationship between groups or individuals, or as part of a dowry when negotiating for a wife. Tabua also often accompanied a man to his grave, as Fijians believed they aided the spirit on its hazardous journey to the afterworld. Some tabua are from sperm whales stranded in Fiji, but many were acquired during trade with Tonga, which had a supply from subsistence whaling that occurred there in the twentieth century.

FASCINATING FACT

Although tabua (ceremonial whale teeth) are often attached to a cord, they are never worn and traditionally were not allowed to be bought or sold.

A Fijian man wearing a wāseisei, a whale-tooth necklace indicating high status.

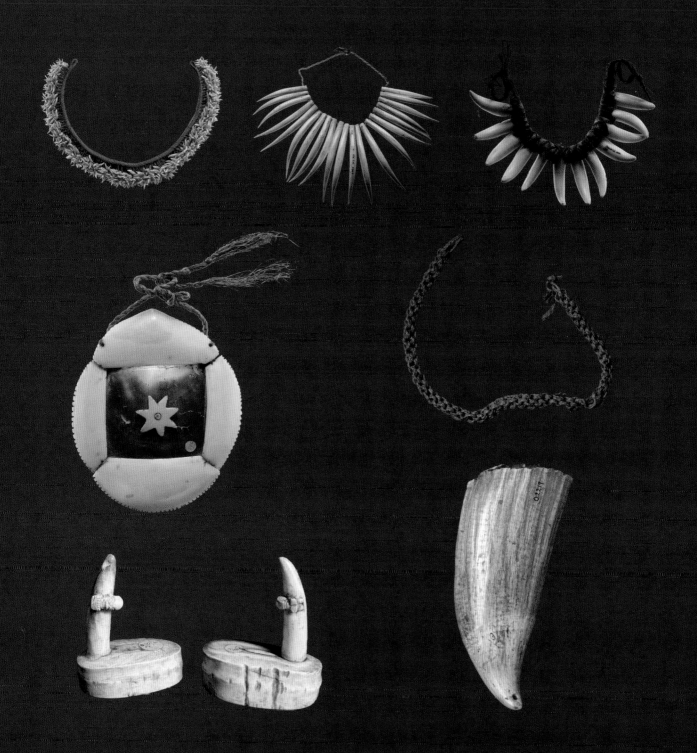

Clockwise from top left: peue ei or peue koi'o (porpoise-tooth crowns) were made from dolphin teeth and glass beads that were strung on a coconut-fibre cord. It would have taken at least five to seven dolphins to make one peue ei headdress, many of which contained up to 1500 teeth.

Wāseisei necklaces were made from split whale teeth and were especially rare because of the great skill required to make them.

Sisi necklaces were made from whale teeth with a hole drilled at their base and coconut-fibre cord threaded through them.

Tabua (ceremonial whale teeth), are always made from sperm whale teeth and are sometimes oiled to darken them.

Ha'akai (ear ornaments) were carved from a single whale tooth. The spike end passed through the ear lobe and the disc faced forward. Ha'akai were highly valued family heirlooms that were passed down from generation to generation through the female line.

This civavonovono (breast plate) displays a whale-tooth star lashed with plant fibre to a pearl-shell plate.

Whale ivory was also used in the Pacific Islands to decorate breastplates called civavonovono, which were worn by chiefs or men of high rank. Civavonovono were usually made by Samoan or Tongan craftsmen using whale ivory, split pearl shell, plant fibre and tapa (bark cloth).

In addition to occasionally hunting whales and dolphins, or utilising stranded individuals, Pacific Islanders had live encounters with them at sea. As the early Polynesian explorers travelled between and beyond their island homes, a number of stories arose that related to the inhabitants in the waters around them. In these tales, whales or dolphins are often symbolised as figures of great power or royalty, and they often guide and protect boats at sea.

TINIRAU AND TUTUNUI

The legend of Chief Tinirau and his pet whale, Tutunui, has been told throughout the South Pacific. The details vary slightly from place to place, but the theme remains the same and reflects the value of whales as friends or guardians, along with their importance as a source of food and bone. The legend is said by some to be the origin of kōhuru, or treachery. In it, Tinirau is often referred to as the son of Tangaroa, the god of the sea; in other legends, Tinirau is regarded as the ancestor of whales or the guardian of all sea creatures. The legend goes as follows.

Following the difficult birth of his son, Chief Tinirau travelled from his island home, Te Motutapu-o-Tinirau, in order to seek the services of the high priest Kae to bless his son. After the ceremony, Tinirau summoned his pet whale, Tutunui, and cut off a small piece of the whale's flesh to give to Kae as a payment for his services. Tinirau offered to carry Kae back to his own island by canoe, but Kae asked if he could ride home on the whale's back instead.

Tinirau reluctantly agreed to lend Kae his dearly beloved whale Tutunui but told Kae that when they neared the shore and the whale shook himself, Kae must disembark so that Tutunui could swim safely back home. Kae, however, remembering how sweet the whale's flesh had tasted, treacherously caused Tutunui to strand, and then proceeded to roast and consume the animal. When the smell of the roasting meat drifted back to Tinirau's island, he realised what had happened. Tinirau was both saddened and outraged, and ultimately captured and killed Kae in revenge for the death of his beloved pet.

Tangaroa ono, a 1996 lithograph of Tangaroa by Michel Tuffery. Tinirau is often referred to as the son of Tangaroa, god of the sea.

MĀORI CONNECTIONS

Ko ngā tohorā ō tātou Tīpuna   *Whales are our ancestors*
Ō tātou kaitiaki   *Our guardians*
Ō tātou tipua!   *Our spiritual guides!*

At some point around or after AD 1200, islanders living in the southern Cook Islands and French Polynesia's Society group set off southwest on several new voyages of discovery. There is no way of knowing exactly what prompted these voyagers to go in search of new land, but whatever the motivation their journeys led to the first landfalls on New Zealand shores.

On their voyages of discovery, the Polynesians used the stars for navigation, but they also looked at other clues to aid them on their journey. The sea itself provided useful information: the colour of the water, types of fish, land debris and particular types of swell patterns all indicated the presence of solid ground. The colour, shape and movement of cloud formations would also have been observed – large areas of land are often covered with convection clouds, which are higher, thicker, darker and slower moving than clouds over the sea. And the navigators would have looked out for migrating birds and whales. In pre-whaling times, thousands of whales such as humpbacks would have migrated annually between the South Pacific and colder waters to the south, and many of the humpbacks would have passed by New Zealand's coastline during their spring journey to their feeding grounds in Antarctica. It's not hard to imagine how the first waka (canoes) could have followed in the 'pathway of the whales', which would have led them to New Zealand. While the culture and society of the initial settlers remained rooted in their ancestral past, there were also changes that occurred as the people adapted to a new environment and developed their own distinctive culture. Ultimately, these people became known as Māori.

TANIWHA

Māori believe taniwha are spiritual creatures that live in oceans, inland waters, dens or dark caves. This bone carving is by Norman Clark. There are many oral histories that tell of taniwha acting as guardians of ancestral waka and then settling down in Aotearoa with the crew of the vessel they had escorted, including this account in Margaret Orbell's *The Illustrated Encyclopedia of Māori Myth and Legend* (1995):

> Tutara-kauika is an ocean taniwha in the form of a whale who protects the people with whom he is associated. In one tradition, he accompanied the waka *Takitimu* to Aotearoa and then remained in the sea as the guardian of the people. In another tradition, he escorted Pou-rangahua's waka, a frail craft that was constructed of bark and albatross skin. He may be accompanied by another companion, a taniwha named Te Wehenga-kauika. Occasionally he leads a group of taniwha, all in the form of whales.

WHALE GUARDIANS

There are many traditional Māori stories and songs that tell of whales acting as kaitiaki, or guardians, guiding and sheltering the waka as they travelled to Aotearoa. A verse from an ancient lullaby composed by the tohunga (priest) Tuhotoariki for his grandson Tuteremoana tells of paikea (whales) calming the waves in treacherous waters. A waka following in the wake of whales would have been surrounded by spray from the whale's blows. Rainbows, created by misty blows in the sunlight, may have given rise to the allusion in the lullaby that the waka were following the pathway of the kahukura, or rainbow:

Hara mai, e tama! E piki ki runga o Hikurangi,
o Aorangi;

He ingoa ia nō Hikurangi mai i tawhiti,

Nā ō kau i tapa.

E huri tō aroaro ki Paraweranui,
ki Tahumākakanui,

Ko te ara tēnā i whakaterea mai ai ōu tīpuna

E te kauika tangaroa, te urunga tapu o Paikea,

Ka takoto i konei te ara moana ki Hāruatai,

Ka tupea ki muri ko Taiwhakahuka,

Ka takoto te ara o Kahukura ki uta,

Ka tūpātia ki a Hine-makohu-rangi

Ka patua i konei te hiringa moana,

Te wharenga moana.

Come, o son, and ascend the peaks
of Hikurangi, of Aorangi;

'tis a name from Hawaiki, from afar,

So called by your forebears.

Turn you towards the mighty northerly blast
and the great blistering easterly wind;

That was the course upon which your
ancestors voyaged hither

Upon the deep sea school of whales,
steered by the sacred ritual of Paikea,

Which becalmed the sea-ways across
the billowing ocean

Leaving in their wake the flying sea spray,

Becalmed was the pathway of the rainbow,

Screened off by Hine, the maid of the heavenly mist,

Subdued too, were the billowing seas

And the curling waves.

WHALE RIDERS

In addition to Māori stories of whales acting as guides or guardians, there are also many oral histories telling of whale riders – people who arrived in New Zealand on the back of a whale or who were carried to safety by a whale when they were in trouble at sea. One of the most famous whale-riding legends is that of Paikea, told here by Hone Taumaunu of Ngāti Konohi iwi:

> The ancestors of most Māori tribes in Aotearoa arrived aboard great migratory canoes. For the people of Whāngārā-mai-tawhiti, however, their ancestor Paikea arrived in this country on the back of a whale. The whole adventure in my opinion is a magical journey of a person, or spirit, or god, who was able to traverse the Pacific Ocean and arrive here physically as a man on a whale. The story begins back in Hawaiki, the spiritual homeland of Māori.
>
> Paikea and his half-brother, Ruatapu, had the same father but different mothers. Their father favoured the first-born son, Paikea. Ruatapu decided that he could offset his shame and embarrassment by killing his brother, and so he invited Paikea and all the other male first-born children to go fishing. Far out at sea, Ruatapu caused the fishing canoe they were in to sink. Ruatapu, Paikea and the other men aboard found themselves in the water. Ruatapu swam among the men, killing each of them with well-timed blows from his paddle, until only Paikea was left. Paikea said, 'Haere mai tohorā, hoki mai ki te manaaki i a au. ('Whales come, I need your help.').
>
> Soon the sea was smoking and blue with tohorā. Paikea picked the youngest and most intelligent-looking tohorā and said, 'Take me home.' The whale didn't know where home was, so they searched around the Pacific. Both Paikea and the whale were horribly lost. They kept moving south and, whether they used the stars, whether they used the migratory birds, whether they used whatever, they kept moving and eventually came here to New Zealand.
>
> So in our opinion, Paikea establishes a link for us with the Pacific, and we perpetuate that genealogy and tradition, which occurred in the Pacific, here in Whāngārā today.

Another legend of a whale rider who came to New Zealand on the back of a whale is recounted by the Kawerau-a-Maki iwi (tribe) and tells of a tohunga (priest) called Rakataura or Hape-ki-tūārangi. The story alludes to conflict in the original homeland as the motivation that led the ancestors to set off in search of a new land, and it also highlights both the traditional and on-going relationships between Māori and whales. The story is told here by Te Warena Taua of Te Kawerau-a-Maki:

This is a story of how my people arrived here in Aotearoa from our homeland, which we called Hawaiki, but specifically the island known as Rangiātea.

Our ancestor's name was Hape-ki-tūārangi. Hape was a tohunga, or priest, and the chief of the village at the time was Hoturoa. There were tensions on the island. The inhabitants were fighting amongst themselves. Hoturoa and his people wanted to leave Rangiātea to find a new home and peace, so they built a canoe. Hoturoa had a beautiful daughter named Kahukeke. Our ancestor Hape favoured Kahukeke, but Hoturoa wanted his daughter to marry some other handsome guy, not this ancestor of ours who had club feet. Hoturoa made a cunning plan – when he left the island on his canoe, he left Hape, the old tohunga, behind.

When the old man found out that this had happened, he cried. He asked for help from Tangaroa, the god of the ocean. In the morning a whale arrived. Hape climbed on the back of this whale and made his journey along the same route that the canoe had taken. Hape was a great tohunga and he used his magical powers so that he and the whale arrived in Aotearoa before Hoturoa's canoe.

According to custom, the land became Hape-ki-tūārangi's because he was the first man to have

Hone Taumaunu of Ngāti Konohi.

Robyn Kahukiwa's Paikea, *1993.*

Whangara, near Gisborne on the east coast of the North Island, is the place where Paikea settled and where his descendants live to this day. Te Motu-o-Paikea, an island just off Whangara, is said to be the fossilised remains of the whale that Paikea rode to Aotearoa.

This carving of Paikea was made by Pine Taiapa and sits at the roof apex of Whitireia, the meeting house at Whangara's marae, built in 1937.

walked it. He turned to the tohorā, the whale, and said 'E hoki, e hoki ki te kāinga' ('Your job is now finished, go back to our homeland'). The whale refused to go. The whale and Hape had developed a relationship, a bond, as they travelled over the waters, over the waves. The whale would not go.

On the beaches here in front of Te Tokatū-a-Hape, the whale died. We have been here 32 generations back to our ancestor Hape. Today, we still live here, on the land that was first founded by our ancestor.

The great whale died on the shores of Manukau Harbour and his mauri (life force) was invested in a whale-shaped rock. This mauri stone continues to be an important connector between the Kawerau-a-Maki people, their original homeland and the great whale that carried Hape-ki-tūārangi to New Zealand.

WHALE PLACES

There continues to be debate about the identity of the first discoverers of New Zealand and the place where the initial landing occurred. The oral histories of Māori include stories of canoes landing in different parts of the country and scientific research bears this out, indicating that a number of canoes arrived in groups and that these craft came from different eastern Polynesian islands. One of the predominant Polynesians credited with the discovery of Aotearoa is Kupe, a fisherman who lived in Hawaiki, the Māori ancestral homeland (thought to be either Raiatea in the Society Islands or Rarotonga in the Cook Islands). Kupe was having trouble with a large octopus, Te Wheke-o-Muturangi, that kept stealing bait from his fishing hook, so one day he set off in pursuit of the giant and chased it all the way to the North Island, eventually killing it in Cook Strait.

Many places in New Zealand bear names associated with Kupe and his crew; some were named by Kupe himself, while others were named to commemorate his presence. One of the traditional whale stories relating to Kupe has to do with the creation of Te Ara-a-Kewa, otherwise known as Foveaux Strait, which separates the South Island from Stewart Island/Rakiura. The story was told by Harold Ashwell, a kaumātua from the Ngāi Tahu iwi in Bluff at the bottom of the South Island, and is relayed here by Kukupa Tirikātene of Ngāi Tahu iwi:

When Kupe the explorer arrived at the bottom of the South Island, he landed at Te Waewae Bay, but the pathway to Stewart Island was blocked by a big sandbank.

Te Warena Taua of Te Kawerau-a-Maki.

FASCINATING FACT

Whangara's Paikea tradition inspired Witi Ihimaera's famous 1987 novel *Whale Rider* and the subsequent 2002 award-winning feature film of the same name.

Kupe went ashore and climbed the hill called Omawero, whose name now is Pahia Hill, and from his vantage point he espied the sea on the other side of the sandbank. Because he did not want to travel via South Cape, he prayed to Kiwa, the atua [god] of the fish, to come to his aid.

Kiwa agreed to assist Kupe and commanded Kewa, the whale, to dig a channel through the sandbank. Kewa set to his task with a will! He bit and chewed his way through the bank until there was a clear passage between the South Island and Stewart Island. When Kewa was clearing the way with his mouth, crumbs fell from his lips and these became the small islands that dot the surface of the strait. As he was chewing away, a tooth loosened and he spat it out. It flew through the air and landed in the ocean far to the west. His tooth still stands there tall and straight, the tooth of Kewa, named Hautere [Solander Island].

The crumbs that fell from Kewa's lips were all given names too, but the most famous name of all was bestowed on the long stretch of water separating the two islands, Te Ara-a-Kewa, the Pathway of the Whale.

The stories of whales were and still are prevalent throughout the country, some going back many generations while others are more recent. Māori named many other places in New Zealand in order to commemorate specific events and stories. Among other things, the names referred to significant stranding events, navigational pathways and important journeys.

One of the early landing sites of the Polynesian settlers was a bay in the Bay of Plenty. When the occupants of the first canoes arrived on the beach, they discovered

This tukutuku panel (woven lattice-patterned panel) features Kupe on the right side of the panel, with his daughters Matiu and Mākaro at the bottom. On the left is Māui, and below him the two taniwha of Wellington Harbour, Whātaitai and Ngake.

Te Ara a Kewa, the pathway of the right whale, or Foveaux Strait.

Three hills in Welcome Bay are known as Ngā Tohorā-e-toru (The Three Whales).

a whale stranded on the sand, so they named the bay Whangaparaoa (Bay of Whales). The stranded whale is said to have caused a huge squabble between the occupants from the *Tainui* and those from the *Te Arawa*. According to some stories, the *Tainui* arrived first and its occupants laid claim to the bay and the whale by placing a piece of rope over the whale. They were tricked, however, by the slightly later arrivals in *Te Arawa*. The captain of *Te Arawa*, Tamatekapua, burnt a new rope to make it look aged and then passed it under the one laid by the *Tainui* crew so that it appeared as if the *Te Arawa* occupants had already laid claim to both the whale and the bay itself. In another version, Tamatekapua produced a whalebone weapon, which he claimed he had already made from the stranded whale, thereby proving he had landed in the bay prior to the arrival of the *Tainui*.

Many of the landforms along the eastern coast of the North Island have traditional names relating to whales. In the Bay of Plenty, known to Māori as Te Moana-a-Toi (the Sea of Toi) after one of the region's first Māori voyagers, Toitetuatahi, whales and dolphins were and still are prolific. The bay stretches from Tauranga to East Cape and a number of iwi settled along its coastline to take advantage of the favourable living conditions; crops flourished in the mild climate and the sea yielded a bountiful supply of fish and shellfish. The Māori people in

Tauranga hold many oral histories about whales in Te Moana-a-Toi. Some histories speak of whales that arrived in Welcome Bay to give birth, and others of whales that came into the bay to scratch their backs in the shallow geothermal sands of the inner harbour.

Ngā Tohorā-e-toru (The Three Whales) are three hills sacred to the tribes Ngāti Pūkenga, Ngā Pōtiki and Ngāti Hē, who believe them to be a family of stranded whales. The story goes that a mother and calf ventured too far into the harbour and were unable to return to the ocean. The father heard their cries of distress and came to their aid, becoming stranded himself. Now the family members remain together for all time. Māori believe that whales stranding in the area are drawn to a spring at the base of the mother's hill. The spring, which sometimes runs milky white, is called Te Waitī-o-te-tohorā, the Milk of the Whales.

Another well-known Bay of Plenty legend is from the Ngāti Awa iwi. The story refers to the tohunga Te Tahi-o-te-rangi who lived in the area known as Whakatane. The story is told here by Ramari Stewart, of Ngāti Awa descent:

> Te Tahi-o-te-rangi had become unpopular with his people, who believed he was misusing his spiritual powers. The people wanted to get rid of Te Tahi but were afraid to kill him, so they decided on a plan to exile him on an offshore island. Te Tahi accompanied a shark-fishing or birding voyage to Whakaari/White Island and made landfall to perform traditional seasonal rites on the island. While he was performing the rituals the voyagers sailed away, leaving Te Tahi behind. Recognising his plight, Te Tahi invoked the aid of the whales to take him back to the mainland. A whale named Tūtarakauika appeared and carried Te Tahi back to land. The whale suggested vengeance and wanted to destroy the voyagers' waka and drown them. Te Tahi replied, 'Waiho! Mā te whakamā e patu' ('No, leave them! Let their shame be their punishment'). He instructed the whale to take a wide berth, so they would not be seen. As the voyagers approached the Whakatane River mouth they became aware that a lone figure standing on Rukapō (a rock) was Te Tahi.

When Te Tahi passed away he was interred at Ōpuru beside the Rangitāiki River. The urupa (buriel site) was later washed out during a deluge. This is in keeping with the notion that a pod of whales came up the river and carried his remains out to sea. There is a history of Te Tahi's descendants appealing to him during times of trouble on the ocean and being rescued by whales.

Moutohora, or Whale Island, off the Whakatane coast was named for its shape, which resembles a whale – the large

Te Ara-a-Paikea, the Pathway of the Whale, is now attached to the mainland by sand dunes.

Today, Moutohora, or Whale Island, is a sanctuary for indigenous forest and wildlife. Dolphins continue to be sighted swimming and playing around the island, along with resident and migrating whale species.

front part of the island is the whale's head, which slopes down to its back and then flattens out into its tail. The descendants of the *Mataatua* waka, one of the early Māori ancestral canoes, occupied the island, which was full of kai (food) such as tītī (grey-faced petrels, or muttonbirds). An abundant amount of kaimoana (seafood) was easily accessed from the island as well, and dolphins and whales were also prevalent in the surrounding waters.

Around Mahia Peninsula, south of the Bay of Plenty, were other popular settlement sites. The local tribe, Ngāti Rongomaiwahine, has a long association with whales. Their oral histories tell how Te Ara-a-Paikea (the Pathway of the Whale), a whale-shaped hill on the isthmus of Mahia Peninsula was once an island. According to the iwi, humpback whales used to pass through a channel that lay between the island and the peninsula. The area around Mahia has always been noted for large numbers of both live and stranded whales.

WHALES AS FOOD

When the first Polynesian voyagers arrived in New Zealand, they found a land of plenty and the seas filled with kaimoana. The new migrants caught birds, fished and gathered seafood to supplement their diets, and there is evidence that some dolphins and, possibly, some smaller whale species were hunted. Within a few

hundred years, most of the seal rookeries in the North Island were depleted and the main populations of moa had been exterminated. Stranded whales remained a welcome source of protein, especially as other sources of meat became more scarce.

Māori learned early on that certain locations were predisposed to whale strandings, and they settled in these areas to take advantage of the koha (gift) from the sea provided to them when a whale or whales came ashore. Every part of the stranded animals was utilised. Whale meat was consumed as food and stored in raised pātaka (storehouses), the oil was used to anoint the body, and the bones and teeth were crafted into ornaments, weapons, musical instruments and other treasured objects.

FASCINATING FACT

At the peak of New Zealand's inshore whaling industry in the 1840s, there were nine whaling stations around Mahia Peninsula on the east coast of the North Island.

RAMARI STEWART:

He taonga Tangaroa

I grew up at Ohope Beach, in the tribal area of Ngāti Awa and Te Whānau-ā-Apanui. My traditional Māori values and mātauranga Māori (Māori knowledge) form the foundations of my worldviews.

I've been involved with research on cetaceans since a project in the Bay of Plenty in 1979–82, where I identified a resident population of common dolphins. I later worked at Campbell Island weather station in 1982–83 and in the mid-1990s to conduct surveys on the rare New Zealand southern right whale. In the inhospitable subantarctic winter, I witnessed some unforgettable and spectacular whale behaviour at Northwest Bay. This was the most southern and isolated breeding ground for right whales. It was truly amazing to see these gentle giants occupying such a small bay. It was a sight and sound of winter that my ancestors would once have witnessed around the shallow coastal bays and harbours of mainland New Zealand.

In traditional Māori culture, beached whales were recognised as he taonga Tangaroa or he koha ā Tangaroa, meaning that they were a treasure or a gift from Tangaroa, the god of the ocean. As a 'waste not' culture, our tikanga, our practice, was to utilise beached whales regardless of their state when they came ashore. Freshly beached whales were a very important food source

in the absence of land mammals. Māori ate fish and birds, but whale meat was highly regarded and an essential part of our people's health. As the whale decomposed on the beach, other resources could still be recovered, right through to the last state of a mummified carcass.

I was granted a permit (1980–2000) under the Marine Mammals Protection Act 1978 to approach and research live whales, and to flense and recover biological specimens and skeletal material from dead whales exclusively for the benefit of research institutions. The Act, however, prevented Māori customary access to beached whales. This was eventually challenged in the mid-1990s when a courageous Māori woman first stepped out onto the beach at Otaki and claimed a sperm whale for her tribe under the Treaty of Waitangi.

Today, tribal rūnanga (councils) have established a Treaty partnership with the Crown and most have a marine mammal protocol with the Department of Conservation (DOC), allowing access and extraction of resources from dead stranded whales. Because the people in many areas have lost the expertise in harvesting parts of whales for cultural use, there is, however, a tendency to turn to DOC for the collection.

In terms of preserving our oral histories and our knowledge, we need to maintain our traditional

practices. In recent years, the focus has been
on regaining our customary right to recover
resources from whales. We have a long way to go
to learn again about the recovery, cleaning and
processing of the materials for future use by
our artists and practitioners. Over time, however,
it is hoped that the relationship will strengthen
between those who recover the material and process
it, and those who utilise it.

Since the 1990s I have actively supported the
countrywide revival of Māori customary use of
stranded whale resources, and to this day I
respond to numerous iwi-related whale-stranding
events, including four current projects in
Te Wai Pounamu, the South Island. Today, the
primary target is mostly sperm whale jaws and
teeth for carving, but other resources such as
whale oil for rongoā (Māori healing) are also
recovered where possible.

My response to the people at a hui (social
gathering) in 1997 marking the return of sperm
whale taonga at Kapiti was: 'Only when we have
bathed in the hinu and the toto (the oil and blood)
of the whale will we understand what it is to
wear a whalebone taonga'. The ancestors of today's
whales knew my ancestors, so it is not just the
physical entity of the whale that lies dead on the
beach, but the spiritual entity too. I strongly
believe that it is important that we as a people
learn again to harvest our own taonga in accordance
with traditional principles and to preserve the
mauri (life essence) of each individual whale
within each of the recovered resources.

He niho, he kauae — what good is a tooth, without
the jaw to carry it? Encapsulated within this
whakataukī, a proverb from my Ngāti Awa people,
is a collective world view and a shared commitment
to nurturing our traditions and the natural
environment in which we live.

WHALEBONE AND WHALE-TEETH ORNAMENTS

Items made from whale bone were considered especially precious because they celebrated the nature of the whale, and a whale's ivory teeth were particularly coveted owing to their rarity and beauty. Ornaments made from whale bone and teeth were carved with great skill and most were worn only by people of nobility.

Hei tiki (human figure pendants) were, and still are, highly prized ornaments, worn around the neck of both men and women, and passed down through the generations. They are often named and each has a story to tell. All tiki acquire mana (prestige) and tapu (sacredness) by becoming imbued with the mauri of their successive owners, and become more precious through these associations with tūpuna (ancestors).

Other pendants made from whale bones and teeth were koropepe and rei puta. The koropepe is in the shape of a coiled manaia-headed (beaked) creature and symbolises a tuna (eel) in motion. Koropepe were made from pounamu (greenstone) or bone, and were worn as a neck ornament. Rei puta are whale-tooth pendants (rei means 'ivory' and

puta means 'extracted from a whale's mouth'). The rei puta pictured opposite has carved eyes and was made from a sperm whale tooth cut in half. Rei puta were rare and very prestigious, and were worn by high-ranking men of great importance in both Māori and other South Pacific island cultures.

To Māori, the head is the most sacred part of the body. Men would wind their long hair into a topknot, which was then secured by a heru, an ornamental hair comb. These were normally made of wood, so a whalebone heru, such as the one shown opposite, carved from a single piece of bone, would have been especially prized. Some combs had simple designs, while others were highly ornate.

Aurei are curved pins that were fastened to a cloak, sometimes several being attached together when the cloak was worn by a person of importance. The rattling of the aurei might then have signalled the approach of a rangatira (leader or chief). Aurei were made from the teeth of smaller cetacean species, or were carved from wood or pounamu.

Carved items were often stored in a pātaka taonga, a treasure storehouse.

*Clockwise from top left: an aurei (cloak pin) carved from
the tooth of a sperm whale.*

*Because sperm whales were not actively hunted in traditional
Māori society, whale ivory and bone were sourced from beached
whales and rei puta made from whale ivory are relatively rare.
This rei puta (pendant) is attributed to Ngā Puhi.*

*This koropepe (pendant) dates from around 1800 and is made
from sperm whale bone.*

*This tiki (pendant in human form) belonged to the Church
Missionary Society missionary, the Reverend Richard Taylor
(1805–1873) and is believed to have been presented to him at
the signing of the Treaty of Waitangi in 1840.*

*Ornamentation was important to Māori, as shown in this
Sydney Parkinson drawing. The presence of a rei puta,
a whalebone heru (comb), a cloak pinned with an aurei,
ear adornments and moko all signify that this was a very
important person of high rank.*

WHALEBONE WEAPONS

Māori were great warriors who used weapons made from wood, pounamu and larger pieces of whale bone. Whale bone gave nobility to an instrument of warfare and mana to the person who owned and used it.

The art of fighting with a taiaha is as beautiful as it is deadly. Warriors skilled in the use of this long-handled weapon can feint, thrust and parry with great force and economy of movement. One end of the taiaha is intricately carved in the shape of an upoko or head. The upoko has a face on each side with a carved tongue poking out. The taiaha is gripped around the tauri, a collar of feathers or dog's hair, which is attached under the upoko. Mau taiaha, the ancient form of stick fighting, is still taught and practised in New Zealand.

Hoeroa, or whalebone staffs, could be owned only by rangatira. Hoeroa are thought to have been used primarily as a staff of chiefly authority, but they were also used in a manner similar to taiaha or thrown at a fleeing opponent. Some hoeroa were carved or decorated with tufts of animal hair. They were usually fashioned from the jawbone of a sperm whale or from the ribs of a large whale species:

> The strangers were numerous and appeared rich: their Canoes were well carvd and ornamented and they had with them many weapons of patoo patoos [patu] of stone and whale bones which they value very much; they also had ribbs of whales [hoeroa] of which we had often seen imitations in wood carvd and ornamented with tufts of Dogs hair.
>
> Joseph Banks, Endeavour *journal*,
> *August 1768 – July 1771* (26 November 1769)

Patu parāoa are short-handled weapons that were fashioned from the jawbone of sperm whales (parāoa means both 'sperm whale' and 'chief'). Other similar weapons, such as wahaika and kotiate, were also used

in hand-to-hand combat. The wahaika is a single-blade weapon distinguished by having a concave recess in the blade that usually displays a carved human figure. Wahaika were made of wood, stone and whale bone. Kotiate are distinguished by having notches on either side of a broad, flattened blade, and were made of wood and whale bone, but never of stone.

Manahi Rangiriri, Te Arawa chief from Tewera, Rotorua.

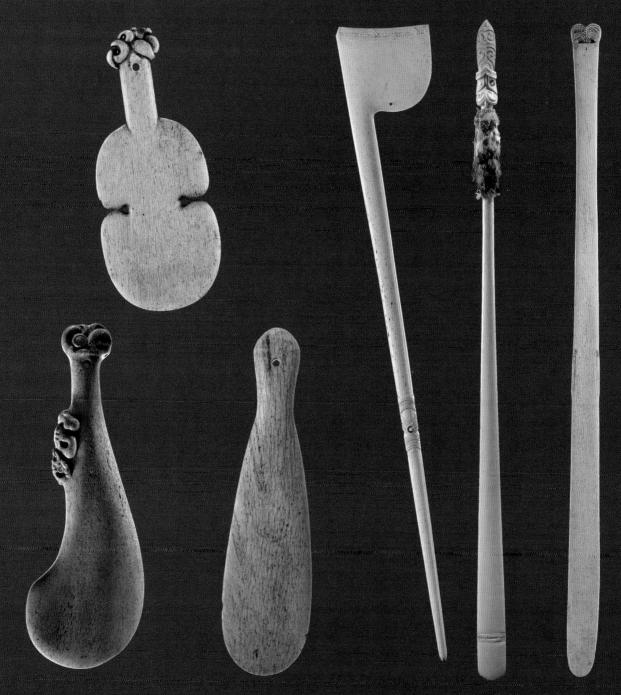

Left to right: this wahaika (short-bladed weapon) is linked to Captain Cook's voyage on the Discovery, which indicates it is at least pre-1777 and therefore pre-European. On one side of the wahaika is a glued note with the following inscription: 'Brought from New Zealand in 1777 by midshipman Burr in the Discovery. Given to his sister in Bath, when an old lady, gave it to Dr. Joseph Hume-Spey of that town, in whose family it remained until October, 1909'.

This kotiate (single-handed striking weapon) features a carved manaia (seahorse) on the handle.

This patu parāoa (hand club) has yellowed with age and dark veins are evident in the bone.

Whalebone tewhatewha (long-handled weapons) were often used by rangatira to direct their warriors' manoeuvres during a battle.

This taiaha (long club fighting staff) was made by Jacob Heberley (1849–1906), the son of famous whaler James 'Worser' Heberley and his wife, Te Wai.

This hoeroa (throwing weapon) is said to have been gifted to Queen Victoria by a Māori chief when he visited England. It was returned to New Zealand in 1948.

WHALEBONE FISHHOOKS

Fishing was essential to Māori, who made a range of matau (fishhooks) and took great care in their design. Their shapes and sizes varied depending on the species of fish being targeted or the fishing technique being used, and they were sometimes made of whale bone, which is stronger than wood. Decoratively carved fishhooks were made to please Tangaroa, the god of the sea, so that he would provide the fisherman with a bigger catch.

WHALEBONE MUSICAL INSTRUMENTS

The first music in Aotearoa was produced on a variety of instruments (taonga puoro) crafted by Māori. Many of these instruments were made of bone, and in earlier times moa, albatross and even human bones were used along with whale bone. Each musical instrument would be seen as unique, with its own particular sound depending on the technique of the person playing it, the type of material used to make it and, in the case of wind instruments, the placement of the holes and size of the bore. Many musical instruments were small and portable so that people could carry them and use them at any time. These included semi-closed flutes, known as nguru, which were sometimes made from sperm whale teeth.

The shape of the nguru is unique to Aotearoa, and the instruments may be played through one nostril or through the mouth. The finely carved ivory flute pictured opposite has three holes, while most nguru have only two. Nguru were usually carved from wood and stone; those made from whale teeth were owned only by tohunga and rangatira. Nguru were played at ceremonies and often in times of great sorrow, such as when a death occurred.

Kōauau, smaller flutes, were sometimes carved from human, moa, albatross or whale bones, and were one of the most common flutes played by Māori, sometimes several being played together to create a chorus. Longer versions of kōauau were known as pōrutu and have a sound that is higher in pitch.

The tumutumu is a rhythmical instrument, again sometimes made from whale bone – the example opposite was crafted from the jawbone of a pilot whale. When struck in different places, the sound it produces varies in pitch and intensity, making it a very versatile instrument that can be played alone or with other types of musical instrument.

WHALEBONE CEREMONIAL OBJECTS

While most items made from whale bone or ivory had a specific use, it appears that some were made as ceremonial objects representing a more practical counterpart. An example of this is the beautiful whale-bone hoe, or paddle, shown opposite. As paddles were usually made from wood, this one was probably ceremonial. It features a carved face on the blade and two carved figures back-to-back at the end of the handle.

The relationship between Māori and whales changed dramatically at the end of the 18th century when European whalers first appeared in New Zealand waters. Previously viewed as koha, or gifts, from the sea, whales soon became a commodity with European financial values. By the early 1800s, Māori men were participating in whaling activities, and by the 1840s whaleboats were being widely used by them and almost half of the whalers in Aotearoa were Māori. The visiting European whalers also had a huge impact on Māori society, forging relationships as they relied on local iwi for food, water, knowledge and protection.

A kōauau (small flute) made by master carver Brian Flintoff.

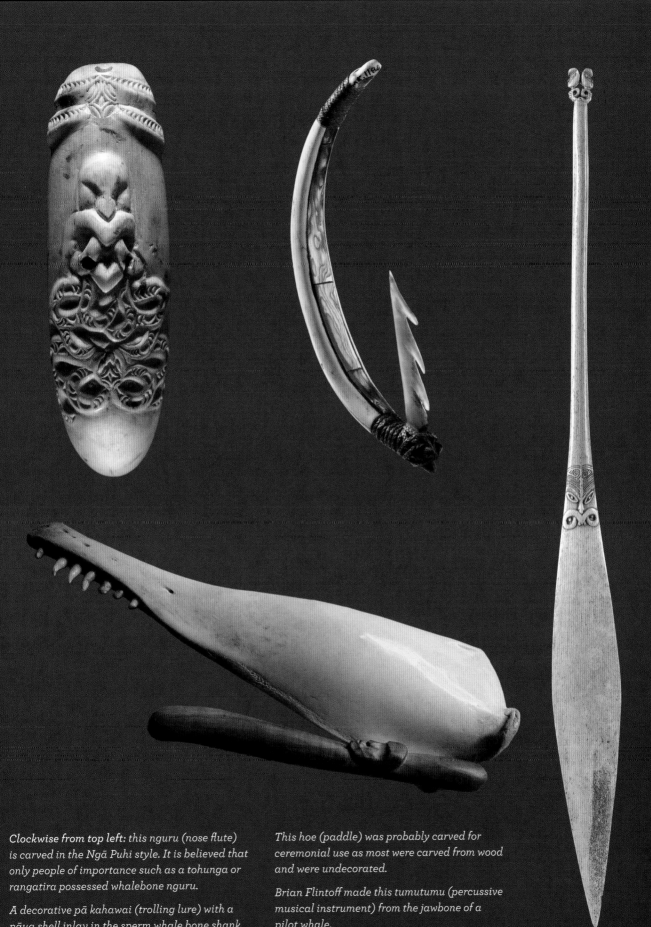

Clockwise from top left: this nguru (nose flute) is carved in the Ngā Puhi style. It is believed that only people of importance such as a tohunga or rangatira possessed whalebone nguru.

A decorative pā kahawai (trolling lure) with a pāua shell inlay in the sperm whale bone shank

This hoe (paddle) was probably carved for ceremonial use as most were carved from wood and were undecorated.

Brian Flintoff made this tumutumu (percussive musical instrument) from the jawbone of a pilot whale.

WHALING

When the last great whale died
no sigh was heard upon the land
but in the heaving of the tide
with every throb, the oceans cried
and cursed the ways of modern man.

Anonymous

THE FIRST WHALERS

It's difficult to pinpoint the exact origins of whaling, but the discovery of a toggling harpoon head at the L'Anse Amour burial site in Labrador, Canada, has led to speculation that whales may have been captured there as early as 5000 BC. Cave drawings in Norway indicate that people may have been hunting whales there by 1800 BC, while evidence from Greenland's west coast suggests the Saqqaq culture started supplementing their diets with baleen whales around 1600–1400 BC. It is probable that this type of subsistence whaling originated in coastal communities that had already benefited from the bodies of stranded whales. It would have been logical, when food was scarce, for such communities to move on to seeking ways of capturing live whales or dolphins and, as time went on and skills were perfected, subsistence whaling would have become a way of life. In some coastal and island communities, such as the Marquesas in the South Pacific (see page 87), people could simply herd dolphins or whales into shallow water. Long-finned pilot whales were hunted in this fashion for centuries in places such as Orkney and Shetland, and there are accounts of whale drives in the Faroe Islands as early as the mid-16th century, a practice that still continues today. Other societies developed ways of going out to sea to capture whales in open waters.

An example of a coastal society that relied on open-water whaling can be seen in the Arctic. The Inuit people in northern Alaska were supplementing their diet with whale meat by AD 400 and by AD 900 whaling was completely integrated into their culture and way of life. Around 1200, these people spread into Canada and Greenland, taking their whaling culture with them.

While the Inuit relied mainly on the capture of bowhead whales, they also expanded their catches to include greys, humpbacks, minkes, belugas and narwhals. The whalers hunted their prey from small skin-covered driftwood boats known as *umiat* (also *umiaq* and *umiak*). This was a particularly hazardous affair, requiring careful teamwork. An Inuit harpoon was composed of a range of animal materials combined with wood and stone. The shaft was often driftwood, and the fore-shaft was made from caribou antler, the finger-rest from walrus ivory, the lashings from caribou sinew, the head from whale bone and the blade from slate. A line made from walrus hide was used to attach the harpoon to a sealskin float. Today, whaling remains an important part of the Inuit culture, as it does in other parts of the world, including Korea and Japan, where subsistence hunting for whales and dolphins from land and from boats also began centuries ago.

A WHALE OF A PROFIT

Although there is evidence of early systematic whaling in the northern seas off Iceland and Norway, the first organised commercial boat whaling, for meat and oil, appears to have been undertaken by the Basques of northern Spain in the 11th century. They targeted the large, slow-swimming northern right whale, now one of the most endangered whale species. In 1984, author and environmentalist Farley Mowat wrote in his book *Sea of Slaughter*, 'Historians have only recently begun to realize ... that it was the Basques who lit the flame that was eventually to consume the mighty hosts of the whale nations.' The Basques had markets throughout Europe and even had consulates in England, Denmark and Holland to encourage sales. The industry spread, and by the early 18th century commercial whalers ranged throughout the northern hemisphere, from the Arctic to the Atlantic, Pacific and Indian oceans. The first northern hemisphere whalers arrived in the southern oceans at the end of the 18th century, and by the 1840s, there were no fewer than 900 whaling ships worldwide, at least 720 of them originating from North America.

WHALING IN AOTEAROA NEW ZEALAND

When James Cook first sailed into New Zealand waters in 1769, he wrote about the vast numbers of whales and seals he encountered and his early charts of the area were illustrated with drawings of these marine mammals. In 1773, the explorer made particular note of large colonies of fur seals in Dusky Sound, Fiordland, which attracted the attention of entrepreneurs in New South Wales, who had almost exterminated the fur seals in Australia's Bass Strait. Sperm whaling was also an attractive option. In these early days of southern hemisphere whaling, baleen whales were seldom targeted for their oil, meat or baleen as the highest prices being paid in Britain and the USA was for spermaceti (oil) obtained from the head of sperm whales. This fine oil was used as fuel for cooking and heating,

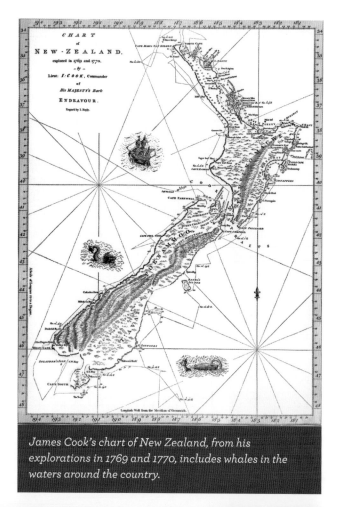

James Cook's chart of New Zealand, from his explorations in 1769 and 1770, includes whales in the waters around the country.

Cutting up a whale, *Wellington, New Zealand, May 26th 1852. Early whalers had to rely on sail power to reach the whaling grounds and once there they relied on manpower, in the form of rowing boats, to chase down, harpoon and capture their quarry.*

The North Cape, New Zealand and Sperm Whale Fishery, *an engraving by Joel Samual Polack, 1838.*

and to light street lamps and homes. By the early 1790s, both sealing and whaling vessels were planning voyages into New Zealand waters.

The whalers were attracted not only by the accounts in Cook's journals but also by the entries in the logbooks of ships carrying convicts from England to Australia. In 1791, Thomas Melvill, master of the *Britannia*, reported, 'within leagues of the shore we saw sperm whales in great plenty; we sailed through different shoals of them from twelve o'clock in the day till after sunset, all around the horizon as far as could be seen from the masthead'. Within 11 days of 'dispatching his live cargo' (convicts) at Port Jackson, New South Wales, Melvill, in the company of another convict ship, the *William and Anne*, headed out to sea to capture sperm whales. The boats managed to kill seven whales, but, owing to a large gale that blew up, they had to cut away five of the whales and returned to port with only one whale each. Undeterred, Melvill continued to go after more whales, although the 'weather was not in his favour'. On 29 November 1791, he wrote to the *Britannia*'s owners in London, proclaiming 'we have the pleasure to say we have killed the first four sperm whales off this coastline'.

It was convict ships such as these that became the first sealing and whaling vessels around New Zealand. While the ships were full of convicts on the leg out to Australia, they needed to return with a cargo to England, and seal pelts and sperm whale oil provided the perfect solution. It wasn't long before some ships began sailing directly to New Zealand for the specific purpose of hunting whales.

In 1801, at least six ships were hunting sperm whales off the northeast coast of New Zealand, and by 1810, there were no fewer than 12 whaleships, most of which were British. The numbers increased yearly as word spread of the large numbers of whales that could be found in the seas around New Zealand. Sperm, southern right and humpback whales were the three main species targeted by whalers during New Zealand's 172-year whaling history.

As more and more ships arrived in New Zealand waters, contact with Māori increased. Aside from supplying the whalers with food, water and local knowledge, Māori also joined the whaling crews, where they soon earned the reputation for being excellent seamen and skilled harpooners, and before long almost all whaleships had Māori crew. The majority of the ships came from America, but there were also vessels from Britain, Denmark,

SPERMACETI

The most valuable sperm whale oil comes from the head of the whale and is known as spermaceti. Spermaceti is of a much higher quality than the oil obtained from the whale's blubber. When heated, it is both smoke- and odour-free, making it the perfect oil to use for lighting, heating and cooking. The fine oil was also used as a high-grade lubricant in delicate textile machinery and watches.

Spermaceti is found in the 'case' of the sperm whale, an elongated barrel-shaped organ that makes up the bulk of the animal's head, and in the 'junk', soft, spongy tissue that is mostly found under the case and is saturated with spermaceti. At body temperature, the oil flows freely as a clear liquid, but as it cools it solidifies to a whitish wax that somewhat resembles sperm, hence the name spermaceti (Latin for 'whale semen'). Once a whale was killed, the valuable oil was easily obtained, as 2000–3000 litres could literally be bucketed straight out of the enormous spermaceti chamber in a male whale's 4–6-metre-long head. In the 12 years between 1804 and 1925, a total of 164,073,915 gallons (621,087,331 litres) of sperm oil (including spermaceti) was landed at ports in the USA, the product of an estimated 262,134 sperm whales killed around the globe.

There was a high demand for smokeless and odour-free candles made from spermaceti in England and the USA, as well as in other European countries and Africa.

Australia and France. To the crewmen of these ships, whaling was an adventuresome way of life that offered an opportunity to see the world, but it was also very hard and they risked their lives for very little monetary reward. The whalers chased their quarry in flimsy boats, and their sperm whale targets often fought back, smashing the boats with their huge tails or diving when harpooned, causing the boats to sink. A whaler's career could end quite suddenly:

> In the year 1804, the ship 'Adonis' being in company with several others, struck a large whale off the coast of New Zealand, which 'stove', or destroyed nine boats before breakfast, and the chase consequently was necessarily given up. After destroying boats belonging to many ships, this whale was at last captured, and many harpoons of the various ships that had from time to time been sent out against him were found sticking in his body. This whale was called 'New Zealand Tom' and the tradition is carefully preserved by whalers.

Thomas Beale, *The Natural History of the Sperm Whale* (1835)

FASCINATING FACT

The closest substitute to spermaceti is oil from the seed of the jojoba shrub (*Simmondsia chinensis*), which is indigenous to the Sonoran Desert. Oil extracted from the seeds has chemical and physical properties that are similar to those of spermaceti.

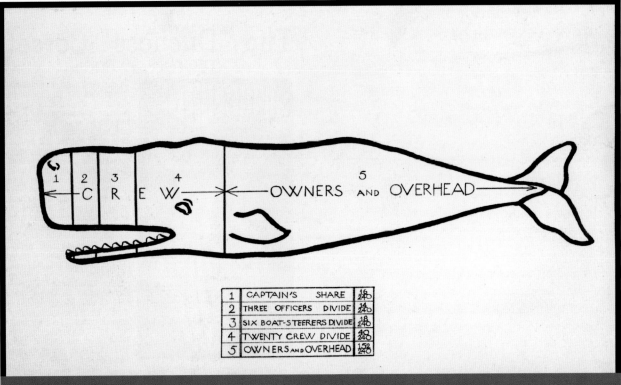

1	CAPTAIN'S SHARE	18/240
2	THREE OFFICERS DIVIDE	14/240
3	SIX BOAT-STEERERS DIVIDE	18/240
4	TWENTY CREW DIVIDE	40/240
5	OWNERS AND OVERHEAD	152/240

'They will rob you, they will use you; worse than any slaves. Before you go a-whaling boys, you best be in your graves,' from 'The Whaleman's Lament', a whaling song from the 1820s–40s.

AMBERGRIS

Oil was not the only valued substance to be obtained from sperm whales. Sometimes whalers got lucky and found a lump of ambergris, also known as grey gold. Ambergris (literally 'grey amber') forms in the large intestine of sperm whales, possibly from irritation caused by squid beaks, and is ultimately passed out when the whale defecates. The waxy substance has a musky odour and is highly valued as a base when making perfumes and also as an aphrodisiac. In his 1881–85 logbook, New Bedford whaler Charles W. Morgan wrote about the *Splendid*, a Dunedin whaling ship that caught a sperm whale containing the largest piece of ambergris ever found: 'The *Splendid* of Dunedin is also noteworthy in that when "spoken" on 20 November 1882, her cargo included 300 barrels of humpback oil, 300 of sperm oil and from a single 60 barrel sperm whale an immense quantity of ambergris weighing 983 lbs [446 kilograms] which was sold for US$125,000 (£62,500).'

FASCINATING FACT

In 1838, Kororareka (now Russell) in the Bay of Islands hosted 54 American whaling ships, along with another 42 from Britain, Australia and France. The port resupplied the ships and provided their crews rest and relaxation in the form of grog shops and brothels. So shocking was the whalers' behaviour, however, that the town gained the nickname 'hell-hole of the Pacific'.

SHORE-BASED WHALING

Up until the 1820s, most of the whalers concentrated on sperm whales, but attention soon turned to the vast numbers of southern right whales (known by the whalers as black whales) that migrated seasonally through Foveaux and Cook straits on their journey to sheltered bays along the east coast of both the North Island and the South Island. These large animals were considered to be the 'right' whale to catch (hence their common name): each whale yielded up to 15 tonnes of oil, they swam slowly and close to shore, they gathered in sheltered bays and, because they contained so much oil, they floated when they were dead.

Although the oil obtained from the right whales was of a lesser quality than spermaceti from the sperm whale, it was still in demand. As a bonus, the whale's mouth contained hundreds of pieces of baleen, known as whalebone, used in upholstery and to make buggy whips, umbrella ribs and the stays in women's corsets.

Soon ships were anchoring in the bays for the specific purpose of harvesting the right whales that had gathered there to give birth. In the late 1820s, the face of whaling in Aotearoa changed dramatically when shore-based whaling was established.

In 1828, Captain John (Jacky) Guard established the first shore whaling station, at Te Awaiti on Arapawa Island in the Marlborough Sounds. Like many of the early whalers in New Zealand, Guard was an ex-convict from Australia who had been a sealer before turning to whaling. He brought his Australian wife, Elizabeth, with him to New Zealand, and their son John, born at Te Awaiti in 1831, was the first Pākehā (European) child to be born in the South Island.

In 1829, far to the south, Captain Peter Williams established Cuttle Cove (Rakituma) in Fiordland's southernmost fiord, Preservation Inlet, also with the purpose of targeting southern right whales. Over the next decade, more than one hundred shore-based stations were established and whaling became New Zealand's major commercial industry. The stations spread from the Bay of Islands in the far north, to Stewart Island/Rakiura in the deep south and to the Chatham Islands in the east. In addition, the bay whalers continued to anchor their ships in the sheltered coves during the right whale calving season. In 1836, in Port Underwood (in the Marlborough Sounds) alone, there were 18 ships at anchor, plus four or five shore stations.

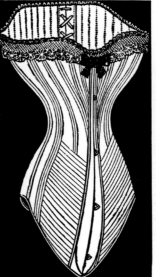

The "Duchess" Corset
(THOMAS'S PATENT)

Is constructed on a graceful model for the present style of dress, the shape being permanently retained by a series of narrow whalebones placed diagonally across the front, gradually curving in and contracting the Corset at the bottom of the busk, whereby the size of the figure is reduced, the outline improved, a permanent support afforded, and a fashionable and elegant appearance secured.

The Corsets are made in Black, White, Dove, Grey. Prices 10s. 6d. and 14s. 6d.

The celebrated TAPER BUSK used in these Corsets is the MOST SUPPLE and COMFORTABLE of ALL BUSKS. In purchasing, it is necessary to see that the name of W. THOMAS is stamped on the Corset.

YOUNG, CARTER, AND OVERALL,
117, 118, WOOD STREET,

Sole Proprietors and Successors to W. THOMAS, late 71, Queen Victoria Street. May be purchased of Drapers and Milliners.

The market for baleen in the 19th century was largely due to a fashion trend. Women coveted hourglass figures, which were achieved by wearing whalebone corsets that made their waists appear smaller. The demand for right whale baleen grew and prices soared.

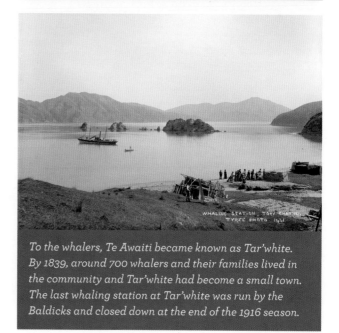

To the whalers, Te Awaiti became known as Tar'white. By 1839, around 700 whalers and their families lived in the community and Tar'white had become a small town. The last whaling station at Tar'white was run by the Baldicks and closed down at the end of the 1916 season.

A right whale being processed on a beach from the New Zealand Herald Weekly, 8 August 1912.

A HARD BUSINESS

Shore whaling in the early days was a hard business. There were no motorised vessels and instead the whaling crew had to row out to capture a whale. Most rowboats had a crew of six men. Four men rowed the boat while the harpooner stood at the ready in the bow with around 200 fathoms (350 metres) of line rolled up behind him. The sixth man was in the stern, and it was his job to steer the boat and position it so that the harpooner could get a good shot at the whale.

A harpooned whale would sometimes drag a boat for hours until the animal became so exhausted that it could be killed with a sharp-pointed lance. The dead whale was then lashed to the boat and towed back to shore, where it was hauled up as far as possible to be processed. The men used cutting spades to carve the blubber off the whale, a technique known as flenching or flensing, and then sliced the blubber into smaller strips so that it could be boiled down in trypots erected on the shoreline. The oil was then drained into large barrels to be collected at a later date.

While some shore whalers had permanent homes erected in the places where they were whaling, others were merely dropped off by a whaling company at the start of a season and then picked up months later at the season's end.

'The boats are sent out at daylight every morning, and when they are so fortunate as to kill a fish it is towed ashore and flinched and boiled up on the beach … The whales are seldom killed nearer than two miles from the harbour, and sometimes seven or eight, and if the tide and wind is against them it is a most laborious business to tow such a huge animal. I have known the boats to be out for 14 hours pulling [on the oars] … Indeed, killing the fish is a trifle in comparison with the getting it in'.

Mr Bell, a Sydney merchant who lived in the Cloudy Bay whaling station for the 1830 season, quoted in Robert McNab's The Old Days of Whaling *(1913).*

HISTORICAL WHALING STATIONS IN NEW ZEALAND

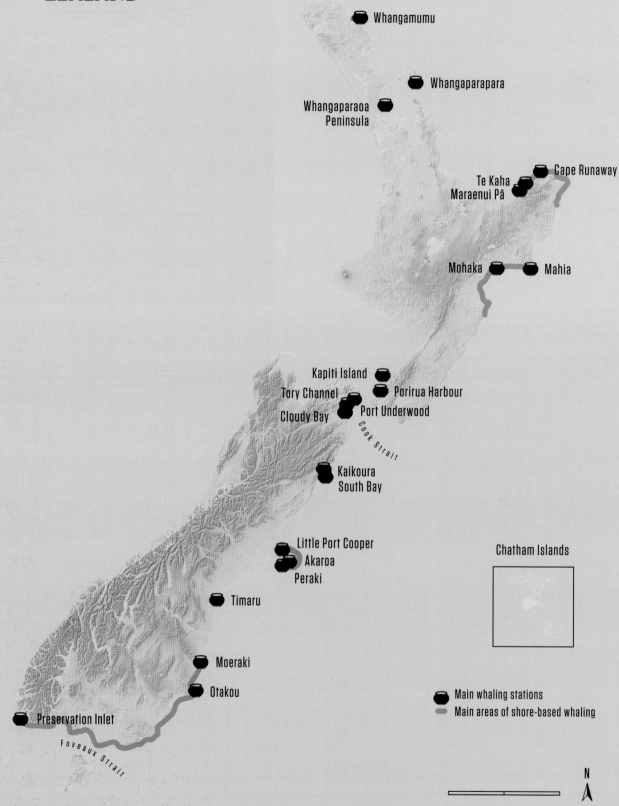

Whangamumu

Whangaparapara

Whangaparaoa
Peninsula

Cape Runaway

Te Kaha
Maraenui Pā

Mohaka

Mahia

Kapiti Island

Porirua Harbour

Tory Channel

Port Underwood

Cloudy Bay

Cook Strait

Kaikoura
South Bay

Little Port Cooper

Akaroa

Peraki

Timaru

Moeraki

Otakou

Preservation Inlet

Foveaux Strait

Chatham Islands

Main whaling stations

Main areas of shore-based whaling

N

Southern right whale

The southern right whale's northern migration started in April and movement continued through September. The whales travelled through Foveaux Strait and up the east coast of the South Island. The slow-moving whales were never far from land and could usually be sighted just off the kelp line within a few kilometres of the shore. Some entered harbours and coves such as Otago, Akaroa, Kaikoura and Cloudy Bay, to shelter and give birth, while others continued on their journey, swimming through Cook Strait and along the coasts of the North Island. Taranaki Bight was a favourite stopping point for the whales and was known to whalers as Mothering Bay. Some right whales deviated from the mainland and journeyed to bays around the Chatham Islands, which led to the establishment there of five shore-based stations.

Humpback whale

Humpback whales migrated at about the same time as the right whales, from their feeding grounds in Antarctic waters to mating and calving grounds in the tropical South Pacific. Humpbacks followed a similar route along the east coast of the South Island, but most were slightly further offshore and did not make long stops on their trip north. Some humpbacks travelled through Cook Strait and deviated to the North Island's west coast, while others continued along the east coast. During the heyday of whaling for southern rights, humpbacks were mainly left in peace. They swam faster, so were harder to catch; they yielded only 4.5–5 tonnes of oil per whale (about a third the amount of the right whales); and their baleen was too short and coarse to be of any great value. However, the whales' peaceful existence was not to last, the humpbacks would become the future targets of the whaling industry.

Shore-based whaling had a profound impact in New Zealand, not only on the whale populations, but also on the lives of Māori. The whalers negotiated with tribal chiefs for the use of land, the right to obtain wood and water, and for protection from attack by the rest of the iwi (tribe) or by invading tribes. Once good relationships were established, the iwi assisted the whalers by growing vegetables and supplying pork or fish. In exchange, Māori received goods, money and work. Many whalers married Māori women, and Māori men worked alongside the European whalers both on the shore and at sea. It wasn't long before almost half the shore whalers were Māori, and some went on to set up their own stations in places such as the Bay of Plenty in the North Island and Otakou on the Otago Peninsula in the South Island.

Many famous whaling families, including the Barretts, Toms, Nortons and Jacksons, and in fact many New Zealanders, can trace their ancestors back to marriages that took place between the first whalers and Māori women. James Heberley, whaler and trader, arrived at Te Awaiti in the Marlborough Sounds in 1830. James, who became known as 'Worser' after constantly declaring there was 'worser weather coming', married Māta Te Nahi (usually known as Te Wai), a prominent member of Te Ātiawa iwi, and together they had two daughters and six sons. Five generations of Heberley men went on to become whalers; fifth-generation whaler Joe Heberley (see page 132) ceased whaling in 1964, more than 130 years after Worser arrived at Te Awaiti.

Whaler and trader James (Worser) Heberley and his wife Te Wai were formally married on 13 December 1841, when the Reverend Samuel Ironsides visited Te Awaiti. During the visit, the reverend also baptised the couple's then three children.

The whalers were often unkempt, colourful characters with nicknames like Geordie Bolts, Long Bob, Flash Bill, Butcher Knott, Gypsy Smith, Fat Jackson, French Jim and Black Peter. They worked hard and played even harder, and it was said the smell of their home-made arrack rum was even viler than the smell of the dead whales. While whaling stations were often rough, odorous places, with the beaches strewn with decomposing whales, flies and scavenging birds, dogs and wild pigs, the whalers' homes were, in general, reportedly kept clean and orderly by their wives:

> I was much interested in observing the life of these rough men and in finding that many generous and noble qualities redeemed their general inclination to vice and lawlessness ... the whaler is hospitable to the extreme and his rough build house is a model of cleanliness and order.

Edward Jerningham Wakefield, *Adventures in New Zealand* (1845).

Killing the goose that laid the golden egg: in a New Zealand Journal *article on 17 September 1842, a French whaling captain was quoted as declaring, 'New Zealand now offers little resource from its latitudes, the black whales, formerly so numerous, have nearly disappeared and there are a great many ships.'*

Thom's Whaling Station, Porerua (sic) in the Marlborough Sounds, engraved around 1843 by Samuel Charles Brees.

FROM ABUNDANCE TO DECLINE

Demand for right whale products remained high and whaling continued to boom in the 1840s, but it was a boom that spelt the end for New Zealand's southern right whales. The unsustainable practice of harpooning both mothers and calves caused the whales to become 'commercially extinct', and by the 1850s, the population was so diminished that it was no longer viable for many shore stations to continue to hunt them and a number were shut down. A few stations operated on a small scale or sporadically when whales happened along, but many whalers were forced to turn to farming or other trades. Among those who had foreseen the collapse was German naturalist Ernst Dieffenbach, who in his 1843 book *Travels in New Zealand* wrote: 'The shore whalers, in hunting the animal in the season when it visits the shallow waters of the coast to bring forth the young, and suckle it in security, have felled the tree to obtain the fruit, and have taken the most certain means of destroying an otherwise profitable and important trade.'

SIR, — I beg to furnish you with the following abstract of the shore whaling parties in Cook's Straite, and on the East Coast of these islands to the southward of the East Cape.

Stations.	Owners.	No. of boats.	Tons of Oil.	Tons of Bone.
Kaikora	Fyfe	4	72	3
"	Wade	2	23	½
Amouri	Wade	3	62	2¼
Waipoppa	Guard	2	18	¾
Cloudy Bay	Dorothy	3	60	2½
"	Thoms	2	20	¾
"	Williams	2	20	¾
QueenC.Sound	Thoms	2	14	½
Mana	Fraser	2	14	½
Kapiti	Gillet	7	140	6
Hawke's Bay	Salmon	4	50	2
"	Perry	4	100	3½
"	Johnson&Co.	3	50	2
Port Cooper	Ames	4	24	1
Ekolaki	Price	3	130	5½
"	Woods	4	100	4¾
Moiraki	Hughes	2	25	1
Waikowaite	J. Jones	2	50	2½
Tyrce	Cheslin	3	55	2½
Toutook	Palmer	2	40	1½
Bluff	Stirling	3	60	2
Jacob's River	Howell	3	60	2
Taranaki	Barrett	2	28	1¼
		68	1215	49

In all 23 stations, giving employment to about 650 men.

W. F.

A record of oil taken from shore stations in the Marlborough Sounds in 1844 from The New Zealand Spectator and Cook's Strait Guardian.

Te Kaha whalers cutting up a right whale killed near Cape Runaway in 1912. Strips of blubber were flensed from the whale and boiled in trypots (right) to obtain their oil. It could take days for the whale to be rendered down and the oil placed in barrels to be shipped and sold in Auckland. The meat was then shared by the community. The vast majority of whales caught by the Te Kaha whalers were southern rights, although other species such as Bryde's whales and humpbacks were also opportunistically taken.

One shore station that managed to continue whaling on an opportunistic basis into the 20th century was run by the Te Whānau-ā-Apanui iwi near Te Kaha in the Bay of Plenty. While most of the whalers had turned to farming, they still maintained a lookout behind the Te Kaha hotel, which had a view of the entire coastline. When a whale was sighted swimming into the bay, a flag was raised or a fire lit to alert the men, who would drop their farming tools and head for their boats. Crew from three nearby stations – Maungaroa, Omaio and Maraenui – worked separately at the start of a chase but would come to each other's aid if needed in order to capture the whale or tow it back to the beach.

In 1842, Robert Fyfe established a shore-based whaling station in Kaikoura. In the autumn of 1843, Fyfe had his first whaling season, starting with four boats and more than 40 men in his employ. Between April and October of that year, the whalers were able to ship off 130 tonnes (123,500 litres) of whale oil. By the end of 1845, Fyfe was the sole owner of two stations. These flourished for a time,

but by 1849 whaling was uneconomic and Fyfe, like so many other whalers, had turned to farming. Robert Fyfe died in 1854 and his cousin, George Fyfe, took over the running of the whaling stations, but they were never a financial success. Kaikoura's shore-based whaling stuttered along using rowboats and hand-held harpoons, but the slow-moving southern right whales were almost gone and the faster-swimming humpbacks were more difficult to catch.

In 1917, two modern whale chasers outfitted with harpoon guns and electronically detonated hand bombs were acquired, and at the South Bay factory, a traction engine was installed to haul up whales, along with a mechanical slicer and large vats to replace the trypots. Even with the modern improvements, however, the whaling continued for only four more years. In 1921, 16 humpbacks and two southern rights were captured, but expenses outweighed profits and shore-based whaling at Kaikoura was abandoned at the end of that season.

Stacks of whale bone outside a cottage in Fyffe Cove, the site of Robert Fyfe's first whaling station.

In 1851, whaler Barney Riley, along with his brother James, set up Rangi-inu-wai station, just south of Kaikoura. Barney had to row out to his whaling lookout, which was situated on the top of a small offshore islet still known today as Rileys Lookout. He was reputedly a good whale spotter, a hard fighter and an even harder drinker, and his Irish wife would often banish him to his rock when he became too inebriated.

When his wife died in the 1860s, Barney placed an advert in the local paper for a replacement: 'Notice to Ladies: Wanted a wife, a widow preferred. A good home and a loving husband. Apply to B. Riley's Fishery.' The ad was repeated several times, but apparently there were no replies, for Barney continued to live alone in his old hut (pictured here) until 1906, when at the age of 75 he was moved into a house of compassion.

As the numbers of right whales around mainland New Zealand and the Chatham Islands plummeted through the 1840s, some whalers started to look further afield. Right whales had been sighted in waters further to the south, but nobody had travelled into subantarctic waters specifically to hunt them. This changed in 1849, when Charles Enderby set sail from Plymouth, England, with 150 settlers in three whaleships, the *Samuel Enderby*, *Fancy* and *Brisk*. It was Enderby's intention to establish a whaling station and a colony in the Auckland Islands, with him as its lieutenant governor. A settlement was built at Port Ross, consisting of 18 dwellings, including a stylish government house, and christened Hardwicke. A gala celebration was held in December 1850 to commemorate the township's first anniversary, but less than two years later the idealistic bubble burst. Too few whales were caught to make the enterprise economic, and the settlers soon grew weary of the harsh conditions and isolated living. In August 1852, Hardwicke was abandoned and Enderby's dream was over.

Half a century later, in 1909, 11 keen whalers from old whaling families in the Marlborough Sounds arrived in North West Bay on Campbell Island, the final frontier in the hunt for southern right whales. The Te Awaiti whalers, mostly made up of Jacksons, Heberleys, Nortons and Toms, brought with them one motorised boat, one rowing boat and two dinghies. Two years later, they were joined by the Cook brothers (see page 122), who set up a much larger operation in Northeast Harbour on the opposite side of the island using their 59-tonne motorised schooner *Huanui* and 44-tonne steam-powered whale chaser *Hananui II*.

Conditions were harsh, however, and the whales were much scarcer than the whalers had hoped. The Cooks' best season was 1912, when they caught a total of 16 right whales and one fin whale, but the following year was an economic failure, so the Cooks left Campbell and returned to whaling

FASCINATING FACT

Campbell Island, 660 kilometres south of mainland New Zealand at a latitude of almost 53°S, is a mere speck in an ocean lashed continuously by the storms of the Furious Fifties. Wind gusts are known to reach 240 kilometres per hour and there is a mere 650 average hours (28 days) of sunlight per year.

full time at Whangamumu. The Te Awaiti men caught only one whale in 1913 and their launch was pounded to pieces in a violent storm, so they too gave up on the whaling and instead returned for two more years to go sealing, quitting the island once and for all in February 1916.

NEW METHODS

With the southern right whale population virtually wiped out and sperm whale numbers dropping to the point where it was becoming uneconomic to go after them, most overseas whaling ships returned home and shore-based whalers turned to other means of earning a living. In the late 19th century, humpback whales had yet to be targeted on a major scale. Although some offshore whaling ships had opportunistically killed humpbacks, the oil from sperm whales was far more valuable and that species was always their preferred catch. Shore-based whalers preferred right whales for the same reason, as well as the fact that humpbacks were faster and more difficult to catch from hand-rowed boats, and usually did not stop to rest and shelter on their migration.

The Te Awaiti whalers did not have trypots for rendering blubber into oil, their aim was to obtain 'whalebone' (baleen) which was worth £1000 or more per tonne. From 1909–11 they caught a total of 31 whales, but numbers dropped after that and the whaling became uneconomic. This photo is from 1912 or 1913.

Whaling south of the Aucklands has proved a complete failure. We understand the weather there was so boisterous during the cruise of the "Earl of Hardwicke" that it was scarcely possible to lower a boat. Whales were plentiful enough, but the difficulties attending the capture of them so great, that scarcely any oil was obtained. This resulted from several causes, two of the principal being that, when killed, the whales sunk; and although they rose in a day or two, the weather prevented the boats from staying by them; and when secured, the whales yielded but little oil, the absence of blubber, in fact, being the cause of their sinking.

The "Earl of Hardwicke" proceeds to Sydney, where, we understand, a last effort is to be made to prosecute the designs of the Company. We fear, however, it is too late to retrieve the losses made, and we predict the speedy wind up of the whole concern. Had the trial been originally made from Sydney, the speculation might have been an excellent one; but the discovery of the gold diggings will in all probability put a stop to the obtaining of crews for whaling voyages, at least for a time; and the attempt to sustain a falling Company, with dispirited shareholders, is, in our opinion, hopeless. It is much to be regretted that a trade of so much national importance as the South Sea whale fishing should have failed from miscalculation; for that it must ultimately succeed we have not a doubt. It cannot be that a trade which is remunerating to the Americans can be unprofitable to the inhabitants of New Zealand and the Australian Colonies, who have all the advantages of being upon the spot.

The failure of the Auckland Islands scheme does not surprise us; indeed, it is no more than we anticipated; for although some portions of the plan are decidedly worth adopting, it was mainly formed on erroneous principles. One of the chief reasons for choosing the Aucklands as a station was to prevent

The demise of the Enderby Settlement at Port Ross recorded in the Otago Witness, *21 August 1852.*

FASCINATING FACT

In December 2010, during the nine-week Campbell Island Bicentennial Expedition, 11 researchers travelled to the island to study its ecology and historical remains. Archaeologists visiting the Cooks' Northeast Harbour station found numerous reminders of the whalers' presence, including three iron trypots.

In the early 1890s, George and Herbert Cook established an operation at the site of an old whaling station at Whangamumu, just south of the Bay of Islands, and their target was the humpback whale. The brothers had noticed that when humpbacks were on their northward migration, they often swam close inshore between the towering cliffs of the mainland and a large, low rock offshore. The Cooks developed an innovative technique to capture the whales by stringing nets made of steel wire and mesh between the large rock and the mainland, which allowed them to trap up to 20 whales in a season. Things were set to get even worse for the humpbacks and other species, however, with the introduction of mechanised whaling.

MECHANISED WHALING

Two innovations in the latter part of the 19th century altered the face of whaling worldwide: steam engines were installed into whaling vessels; and in 1864, Norwegian Svend Foyn devised 'a most wonderful invention', an explosive harpoon gun that could be mounted on the bow of whaling boats. These new methods brought in an era of industrial whaling. Ships could now travel faster and further, and the mounted harpoon guns killed whales with a far greater efficiency than hand-held harpoons thrown from small boats. By the start of the 20th century, the age of modern whaling had begun, an age that was to witness the deaths of more than 2 million whales in the southern hemisphere alone.

This group of Te Awaiti whalers, along with some of their children, are standing in front of Patrick Norton's whaleboat Tangitu, *named after his wife, and James Jackson Jr's boat* Swiftsure. *Jackson was famous for his ability to kill a whale with just one lance. Some of the whalers hold harpoons and lances, while one of the young boys is holding two pieces of right whale baleen. A few years after this photo was taken, the hand-rowed whaleboats were retired, making way for the modern age of whaling that was to follow.*

In Whangamumu, large nets of steel wire and mesh were strung up between a large rock and the mainland to catch whales passing through. Large buoys attached to the nets prevented the whale from diving when it became entangled. The humpback would often thrash its tail violently in the air, trying to free itself from its steel-mesh trap.

A HEROIC OCCUPATION?

It's sometimes difficult in our age of 'save the whales' to think of whaling as a heroic occupation, but it must be remembered that when men around the world first went whaling in the 18th century, they were providing products such as food and fuel that were in some cases necessary for human survival. The men had to be tough, conditions were harsh and death was not uncommon. Offshore whaling ships were dependent on the wind to chase down whales and the men had to be lowered into small hand-rowed boats in rough seas to harpoon large whales that often fought back. Shore-based stations were also reliant on hand-rowed boats to capture whales that were twice the size and many times the weight of their craft. At times, the men had to sever the line between their boat and the whale in order to avoid being towed out to sea.

FASCINATING FACT

Ironically, the first southern right whale sighted in New Zealand in 35 years, on 15 July 1963, was within a few hundred metres of Perano's Tory Channel whaling station. Thankfully for that particular whale, all southern right whales had been given total protection in 1935.

A whaler sitting in the mouth of a dead whale at a whaling station in the Bay of Islands.

In New Zealand, motorised whale catchers were used at the Cook brothers' station in Whangamumu from 1910 until its closure in 1931, and at Kaikoura from 1917 until its closure in 1921, but the predominant mechanised whaling venture took place in Tory Channel, near the entrance to Cook Strait.

In 1911, Joe Perano, 'inspired by an encounter with two humpback whales', founded J.A. Perano & Company. The following year he moved his whaling station from Yellerton Bay (now Te Rua Bay) to Tipi Bay, where it was managed by his brother Charley until 1927. In late 1923, Joseph erected his third and final station in Fishing Bay (now Fishermans Bay) on Arapawa Island, where it remained until whaling ceased in 1964. In the early days of his whaling venture, Joe Perano did not have it all his own way. There was competition from other whalers in the Marlborough Sounds, and fierce rivalries sometimes led to bitter disputes, such as an incident described in the *Marlborough Express* in 1911 (see page 125).

The Peranos were innovative whalers, who used sleek, fast whale chasers with harpoon guns mounted on their bows. Their harpoon gunners included the best in the industry – men such as Trevor Norton, Sid Toms and Charlie Heberley. Norton worked for the Peranos for 27 years and could strike a whale with six out of every seven shots. Humpback whales were the whaler's main

Introducing ...
WHALE STEAK

Though this is the first time that whale steak has been introduced to the New Zealand public, it has been exported overseas from Picton for some years and was subject to inspection by Government inspectors, who passed it fit for human consumption in the same manner as other meats.

Of the 139 whales caught last year, not one was rejected on account of any impurity.

Remember that this is a meat, and not a fish, because a whale is a mammal. The meat is tender and juicy and to obtain the best results in cooking, it should not be cooked too quickly but allowed to simmer to retain the rich juices.

Recommended recipe

Take 3 or 4 medium-sized onions, cut up and brown in pot. Cut 1lb. of WHALE STEAK into cubes, place in pot with browned onions. Add water to cover. Simmer slowly for one and a half hours. Thicken and serve.

This is not the only way of cooking whale steak. You can cook it like any other meats; you may also fry or grill it, but because it is so tender it is not recommended for roasting. Whale steak is nutritious and full of proteins.

Sold Only By
V. A. BARNAO LTD.
397 Main Street   Palmerston North

This advertisement and recipe for whale meat appeared in the Manawatu Evening Standard *in June 1950.*

targets, as southern right whales still hovered on the brink of extinction, but they also opportunistically harpooned a few sperm, blue, sei and fin whales.

Humpback oil provided Perano's company with its main source of income, and its Whekenui Whale Oil was awarded gold medals and diplomas of merit in both 1925 and 1926. They also sold fresh whale meat through established Palmerston North fishmonger V.A. Barnao Ltd, where staff not only gave away a free sample of whale meat with every fish purchase but also provided a recipe (see page 123). Unfortunately for the Peranos, however, the housewives of Palmerston North seemed to prefer the mutton, lamb and beef grazing in their nearby pastures, and whale meat never really caught on.

The years 1959 and 1960 were New Zealand's best humpback whaling seasons. At Great Barrier Island (Aotea Island), a whaling station was set up at Whangaparapara, managed by Charlie Heberley and employing other whalers from the Marlborough Sounds. The station caught 104 whales in 1959 and 135 in 1960. Back in the Marlborough Sounds, 1960 was the Peranos' best ever whaling season. Their first whale was taken on 21 May, by 2 June they had killed 78 whales, and by the middle of June the Fishing Bay processing station was so full it was unable to cope with any new whales and the whalers had to take a few days off so it could catch up. On 7 August 1960, the season ended with a record catch of 226 humpback whales.

THE END OF WHALING IN AOTEAROA NEW ZEALAND

The following year, on 1 June 1961, the Perano family and friends sat down to an anniversary dinner to celebrate 50 years of whaling. By mid-July, the celebration was well and truly over, and many of the whalers had to be laid off owing to a lack of whales. The total catch for the 1961 season was 55 whales, and 1962 proved to be even more depressing, with 24 humpbacks – the lowest in 30 seasons – and three killer whales. The whalers were so desperate they had killed the three orca to find out how much oil they would produce; the amount was negligible. The poor seasons were echoed up north, where the Hauraki Whaling Ltd station on Great Barrier Island closed down at the end of the 1962 season due to a lack of whales. The efforts of large overseas whaling fleets operating in South Pacific and Antarctic waters had finally taken their toll: the populations of humpbacks, like that of the southern right whale, had been decimated.

In 1963, the Tory Channel whalers took only nine humpbacks, but even before the start of that season brothers Gilbert and Joe Perano Jr, sons of Joe Sr, realised they needed to make

The Perano family experimented with many products, including canning whale meat for sale in the UK after the Second World War, and using whale scraps and bone for fertilisers.

Joe Perano and Arthur Heberley stand in the mouth of one of the last remaining southern right whales killed in Tory Channel, in 1918. Two more were taken by whalers in 1927, but from that year on there were no records of any southern right whale sightings along New Zealand's entire mainland coastline until 1963.

a change. In July 1962, they began a survey of sperm whale movements; their first sperm whaling season started in Kaikoura in April 1963 and carried on until the following December, with a final tally of 119. The 1964 season was slightly better, with the whalers taking 129 sperm, one fin and four sei whales, but it all was too late. The price of oil had dropped dramatically, from £85 to £42 per tonne, and it was no longer economically viable to make a living from whaling. The last whale killed by New Zealand whalers was harpooned near Kaikoura by gunner Trevor Norton, the son and grandson of whalers, who had killed up to 1400 whales during his 27-year career. On 21 December 1964, Trevor fired his last shot and New Zealand's whaling history ended with the death of a 13-metre male sperm whale.

FASCINATING FACT

Once it was harpooned, a 50–100 tonne whale could be processed and reduced to 'raw product' in a factory ship in as little as 20 minutes.

THE WHALES.

A DISPUTED PRIZE

Our Picton correspondent writes as follows:--

Word was received in Picton about mid-day yesterday that two whales had been secured in the Sound. The fortunate parties were J. Perano's crew in the Cresent, and Mr James Jackson's crew of sturdy whalers.

Both whales were secured in Tory Channel near Dieffenbach. A special launch went down in the afternoon and took a party of sight-seers to the scene of operations.

Both whales were about 60ft in lenght, and of the hump-back species. It is difficult to estimate their value, but it was estimated at from £80 to £100 each.

Trouble, however, seems to be brewing, and there is every probability of there being another whale case. It seems that there were three crew's out after the two whales. The one captured by Jackson's boat had, it is claimed, been harpooned by one of Norton's crew, which was No. 3 party; but Jackson's redoubtable crew came along and succeeded in getting fast, and they claim the whale as their property, so there is every likelihood of the law being involved to decide if Norton's harpoon, which was the first to strike the whale and was afterwards cut out from the carcase, constituted proof of the whale being his.

A 'disputed prize' recorded in the Marlborough Express, *1911.*

A star whale chaser of the Rosshavet Whaling Company is surrounded by the bodies of four blue whales in Antarctic waters in the early 1920s.

An unidentified Māori whaler lying back on the throat grooves of a whale at a whaling station, Bay of Islands, 1927.

Prior to the age of mechanised whaling, which saw boats venturing into Antarctic waters, many sail whalers also used the waters around the South Pacific islands as their hunting grounds. While they mainly targeted sperm whales, they also took humpbacks and other species when they were available. The waters around Norfolk Island in particular attracted many whalers in the early 1800s, with the islanders resupplying the visiting crews for many years. By the late 1850s, the islanders had established their own whaling industry and whaling from the island continued for more than 100 years. In the mid-1950s, they brought in new motorised equipment to increase their catches of humpbacks, but by 1961 the decline in numbers noted in New Zealand was also being felt at Norfolk Island and whaling ceased there in 1962.

Whaling also occurred around other Pacific islands. Small shore-based whaling stations were set up on some islands, including Vanuatu, and American whalers targeting sperm whales also took humpbacks from the species' breeding grounds around Tonga and New Caledonia. In the mid-1880s, New Zealand whaler Albert Cook set up a bay whaling company in Tonga, and the industry expanded over the years to include 11 small operations employing around 170 people. The whalers chased their prey with sailing and rowing boats, and killed them with hand-held harpoons. Although the number of whales killed was not huge – 10–30 per season – the local whalers were targeting mothers and calves, and elsewhere the same population was being targeted by other whalers. Between July and October 1979, researchers William Dawbin, Martin Cawthorn and Ronald Keller carried out a survey of humpback numbers around Tonga, finding that there were as few as 312 adults and only 15 calves, and that the number of breeding females in particular had been drastically reduced. In 1979, King Tāfa'āhau Tupou IV issued a royal decree banning whaling in Tongan waters.

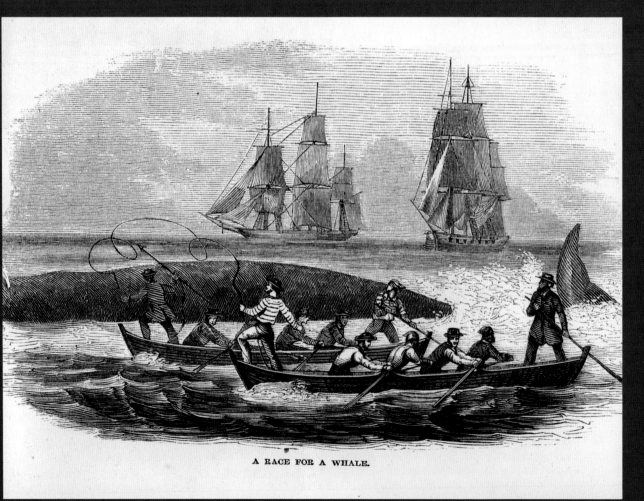

A RACE FOR A WHALE.

CHANGING PERCEPTIONS

During the first 100 years of whaling in New Zealand waters, southern right whales were virtually wiped out by the efforts of shore-based whaling stations and bay whalers. Sperm whales also received a tremendous amount of attention from offshore whaling ships, but although their populations were reduced, they were never diminished to the same degree as the right whales. Humpback whales were targeted from the early 1900s, but the total numbers killed by New Zealand whalers in the 54 years up until whaling ceased in 1964 were fewer than 6000, a small percentage of their overall population. It is estimated that as many as 20,000 humpback whales once migrated past New Zealand to their winter breeding grounds in the South Pacific, but by the mid-1970s – a decade after the end of whaling here – their population was reduced to less than a few hundred individuals. So what had happened?

The humpback whales that pass through New Zealand waters feed in the Ross Sea in Antarctica in summer and then migrate north in winter to breeding grounds around islands in the South Pacific. By the early 1900s, the mechanisation of whaling saw motor-powered whaling ships reaching Antarctic waters, and by 1925 they had been joined by large factory ships that allowed the whalers to remain on the whaling grounds for months at a time. Speedy whale chasers armed with explosive harpoons were now able to chase and kill humpbacks along with other faster swimming whales such as blue, fin and sei, and compressors installed on the ships pumped air into the bodies of the dead whales so that they would remain afloat. The age of modern whaling was now truly underway in the southern hemisphere.

As early as 1930, whale numbers in Antarctic waters were plummeting and alarm bells started ringing. In March 1931, the *Otago Daily Times* reported:

> The slaughter of whales has grown to enormous dimensions and those who have been associated with it since it first began are now showing some concern for the future. The whalers now fear that the life of the industry is rapidly drawing to a close, as the number of ships operating in the Antarctic is so large that the early extermination of the whales is a certainty ... the depredations are such that already the Ross Sea whaling ground is said to have been virtually cleaned out of whales.

In 1931, the League of Nations drew up the Convention for the Regulation of Whaling, which was signed by

This large scrimshawed sperm whale tooth was presented to the Dominion Museum by Captain Solyanik on 8 April 1958. It is engraved with a helicopter flying in front of a whale chaser, depicting the mechanised whaling methods being used at the time.

A Russian whaling fleet anchored in Wellington Harbour in 1958. It was comprised of two large factory ships, a supply ship and 25 whale chasers. A helicopter was also used for spotting prey.

STOP BLOODY WHALING
今すぐ止めよう偽りの学"Scientific"術調査の 残虐捕鯨
⊖ GREENPEACE ☮

22 countries and supposedly came into force in 1935. In reality, however, very little changed.

Whaling slowed down during the Second World War, but post-war food shortages worldwide led to whale meat being viewed as a source of cheap protein, and other products such as margarine, ice cream, cosmetics and fertilisers were also being made from whales. Britain, Japan, Norway, the Netherlands and the Soviet Union, amongst others, sent whaling fleets to the Southern Ocean soon after the war ended. The result was yet another period of indiscriminate slaughter and a further reduction in whale numbers. In 1946, the International Whaling Commission (IWC) was established with the aim of achieving the 'maximum sustainable utilisation of whale stocks'. For many years, countries argued over whale quotas, while the whales themselves continued to be killed in unsustainable numbers.

Although New Zealand was one of the founding members of the IWC, there was not initially any big public outcry here against whaling excesses. For example, when a Russian whaling fleet comprising two large factory ships, a supply ship, 25 whale chasers and helicopters for spotting prey visited Wellington in 1958, they were made welcome and newspapers wrote accounts of their 'adventures on the high seas'. A. Solyanik, the captain of the factory ship *Slava*, proudly stated that in the 1957–58 season alone, about 20 per cent of the Antarctic whale population had been killed by ships from the USSR, Japan, Britain and Norway.

These visiting whaling ships did not restrict themselves to Antarctic waters but also hunted in temperate and tropical seas. In March 1964, the Japanese whaler *Nisshin Maru 3* anchored at Cloudy Bay, having already taken more than 1200 sperm whales from South Pacific waters. A few weeks later, the *Slava* again visited New Zealand. This time the Soviet fleet was whaling off Pegasus Bay near Christchurch, and reputedly 'their smoke was wall to wall across the southern horizon'.

The rapid decline in humpback numbers as a result of hunting at their feeding grounds in Antarctica, along with the associated collapse of the New Zealand whaling industry, pricked the conscience of the general public. Whales in New Zealand had been taken for granted – they were always there, and now suddenly they were gone. The growth of the environmental movement in the early 1970s led to a new and different way of looking at whales, not only here but throughout the world. Images of whale slaughter, along with a heightened awareness of whales as intelligent creatures, sparked a worldwide movement dedicated to saving whales. In New Zealand, organisations such as Project Jonah and Greenpeace led marches to gain support to end the slaughter. A victory was achieved in 1978 when the Marine Mammals Protection Act was passed by Parliament, at last affording protection to all whale, dolphin and seal species in New Zealand waters.

In 1986, some 40 years after its establishment, the IWC placed a moratorium on commercial whaling. For humpback, right, sei, fin and blue whales, it was almost a case of too little, too late. Each of those species had already been reduced to tiny remnants, for some amounting to less than 1 per cent of their pre-whaling numbers.

Southern right whales have been given international protection from commercial whaling since 1935, humpback and blue whales since the mid-1960s, and fin whales since the early 1970s. Today, despite 40–50 years of protection, blue and fin whale populations in the southern hemisphere continue to number in the low thousands. Humpback and southern right whales are beginning to make a steady recovery in the waters of Australia, South Africa and South America, but in New Zealand and the South Pacific, their numbers remain low. For many years, scientists wondered why some species have been so slow to recover. Part of that question was answered in 1994 with the discovery that the USSR had been engaged in a systematic falsification of their catch data from 1947 up until the early 1980s.

The Soviet factory ship Slava *alone processed 2120 more whales than it reported during the 1958–59 whaling season.*

WHALE CATCHES IN THE SOUTHERN HEMISPHERE, 1900–2005

WHALE SPECIES	1900–2005 CATCH TOTAL
Blue	362,770
Bryde's	7,881
Fin	725,331
Humpback	213,245
Minke	119,415
Sei	203,843
Southern right	4,424
Sperm	405,898
Other	11,835
Total	**2,054,642**

The chart above shows the number of known whale catches in the southern hemisphere from 1904 to 1983, based on figures reported in whalers' logbooks. The figure for southern right whales is relatively low as the vast majority of their population had been wiped out almost 100 years before the IWC was even formed. (Source: IWC, *Encyclopedia of Marine Mammals*, 2nd edition, (William F. Perrin, Bernd Würsig, J.G.M. Thewissen, editors), 'Southern Hemisphere Catch Totals: 1900–2005', 2009)

WHALE CHEATS

In 1994, Russian scientist Alexey Yablokov published a paper in the journal *Nature* that sent shockwaves throughout the scientific community and the IWC. The Yablokov report, titled 'Validity of Whaling Data', described the illegal whaling that had occurred on Soviet whaling vessels for almost three decades. The following year, Yablokov published 'Soviet Antarctic Whaling Data', which gave a detailed account of the actual numbers of whales that had been taken by four Russian factory ships between 1947 and 1972, these figures varying greatly from those previously reported to the Bureau of International Whale Statistics. For years, the fleet had returned home with their holds crammed full of illegal catches of protected species – catches that were never declared. The real numbers were hidden from the IWC, to whom false information was officially reported. KGB officers on board the vessels kept the true tally and ensured the crew did not leak information about their cargo when the fleet called into ports.

After the collapse of the USSR in the early 1990s, Yablokov formed a committee with other Russian scientists who had worked on board some of the vessels and together they searched the KGB files to compile lists of the true numbers and species of whales that had been killed. In some cases, catches of unprotected species were over-reported in order to disguise takes of protected species. Three of the most shocking figures relate to the actual numbers of humpback, pygmy blue and southern right whales that were killed, protected species that today are still struggling to recover from the effects of whaling. The figures of illegal southern right whale catches include an estimated 250–288 taken from around the Auckland Islands in the late 1950s. Without this illegal catch, New Zealand's southern right whale population would now be three to four times its current population of 1300–2400 individuals.

An obvious implication of the reports compiled by Yablokov, as well as other papers that have since been presented by scientists who worked on board the whaling ships, is that if four whaling ships had falsified data, it is likely other ships had done the same. Sadly, we will probably never know the real number of whales killed during that industrialised era of whaling.

Although New Zealand was a whaling nation for more than 130 years, it is now one of the leaders of international efforts to oppose commercial whaling. At the 1994 IWC meeting, New Zealand, along with other countries, lobbied for the establishment of a whale sanctuary in Antarctic waters. Yablokov's paper, presented at that same meeting,

ILLEGAL SOVIET WHALING, 1947–86		
WHALE SPECIES	**REPORTED CATCH**	**ACTUAL CATCH**
Blue	3651	13,035
Bryde's	19	1468
Fin	52,931	44,960
Humpback	2710	48,721
Minke	17,079	49,905
Sei	33,001	59,327
Southern right	4	3368
Sperm	74,834	116,147
Other	1539	1405
Totals	**185,768**	**338,336**

In 2010, a new report was released with updated figures on the illegal Soviet whaling that had occurred in the southern hemisphere in 1947–86. These new figures, when combined with the illegal catches in the northern hemisphere, show that more than 180,000 whales were illegally killed worldwide by the Soviet whaling fleet alone. Source: C. Allison, 'IWC Summary Catch Database Version 5.0' (1 October 2010)

FASCINATING FACT

In the 1950s, the Soviet fleet contained the largest factory ships ever built, the *Sovetskaya Ukraina* and her sister ship the *Sovetskaya Rossiya*. They were each capable of processing 200 small sperm whales and 100 humpbacks or 30—35 pygmy blue whales in a single day.

demonstrated that far more whales had been killed than previously reported. The meeting ended in a victory for conservationists – the Southern Ocean Whale Sanctuary became a reality, giving protection from commercial hunting to all whales in waters below latitude 40°S.

Unfortunately for the minke and fin whales that feed within the Southern Ocean Whale Sanctuary, however, some countries continue to exploit loopholes in the International Convention for the Regulation of Whaling (ICRW) that allow them to grant any of their nationals 'a special permit authorizing that national to kill, take and treat whales for the purposes of scientific research'. Every year, Japanese whaling ships continue to kill whales for 'scientific research'; and every year, people continue to fight for the whales' right to live safely within the Southern Ocean Whale Sanctuary.

It is no longer necessary to kill whales in order to learn more about them, and people around the world now study cetaceans using new and innovative research techniques. Fifth-generation whaler Joe Heberley is an example of the change in attitude towards whales and whaling. He and other ex-whalers now work alongside New Zealand researchers conducting annual humpback whale surveys. The former whalers spot and count humpback whales in the very place where they once hunted them.

Greenpeace activists attempt to hinder the transfer of minke whales to a Japanese factory ship in the Southern Ocean in January 2006.

JOE HEBERLEY:

Born to be a whaler

I was a fifth-generation whaler. My great-great-grandfather James Heberley, better known as 'Worser', came out to New Zealand in 1830 with Captain John Guard and worked at his Te Awaiti shore-based whaling station in Tory Channel, just two bays down from where I live today. John Guard was my mother's great-uncle, so whaling is in my blood from both sides of my family. As a boy, I got excited when my father Charlie (pictured below, left) was whaling. From our home, I'd see the chasers occasionally drive a whale into Tory Channel. I'd watch and dream of the day I'd be a whaler like Dad. I'd go along the beach and 'kill' driftwood 'humpback whales' with sticks.

Whalers lived in a bloodthirsty and dangerous environment. Whaling was a job, a very necessary job. During the war, it was a primary industry; the high-grade oil produced by the whales was used in machinery and the hardening of gun metal, and frozen edible whale meat was sent to hungry people in Britain and Europe. As I got older and I'd see my father heading off every morning for the three-month whaling season, I counted the days until I could join him.

In 1959, Dad was asked to manage the whaling station on Great Barrier Island. I'd just turned 16 and worked as a deck boy my first season (opposite), but the following season I became a gunner on one of the small chasers. For me it was like a sport. When a whale was spotted and 'thar she blows' was called, the adrenalin began to pump. As we raced

to our boats, I felt the way an All Black must feel as he races onto the paddock for a rugby game. At first there were always whales, and then the numbers suddenly dropped. We stopped whaling at Great Barrier in 1962; everyone was devastated. I returned to work as a gunner for the Perano brothers, but whale numbers dwindled even further and the operation finally closed in 1964 and whaling in New Zealand ceased forever.

In 2004, Department of Conservation marine mammal scientists Nadine Bott (née Gibbs) and Simon Childerhouse approached me to see if it would be feasible to set up a lookout in this area to do a count of humpback whales. They wondered if I knew any old whalers who would be keen to join the survey... in one evening I had four; our numbers have now grown to eight. The first four years we spent perched on East Head, working from a tent that opened from the front (opposite, right). It was freezing, windy and often wet, but we didn't care. We were transported back to the halcyon days of whaling that we'd all loved. Our 'chairs' were built on the same principle as our ancestors', so that we could sit 'reverse' style watching through binoculars fastened to a stand fitted on the chair back.

When the two-week surveys were extended to four weeks in 2008, we decided we'd had enough of sitting in a freezing tent, so we built a hut to use as our lookout station. We still work in exactly the same way as the older generations and, in the time-honoured traditions of former whalers, a notch is carved on the back of the chair of the person who spots a whale (once the sighting is confirmed).

For all of us old whalers, what we have been doing over the past eight years has been a privilege. We hunted and killed whales because it was our job. During the surveys, we're all into whaling mode again. We still get that adrenalin rush, but now we are working to save the whales. We are committed conservationists and our aim is to try and help to protect these animals by getting as many sightings as possible. We are against the taking of whales in the Antarctic, and especially against the cry that it is for scientific reasons. And now when the whales are scarce, every whale that we see today is a thrill.

As for me, one of the most rewarding outcomes is sitting next to my sons as well as their wives and my grandchildren, who are continuing the long-standing family relationship with whales.

WHALE RESEARCH

*The living Leviathan has never yet
fully floated himself for his portrait.*

Herman Melville, *Moby-Dick* (1851)

Whales are difficult to study. They live in a world that is alien to us, giving glimpses of themselves when they surface, but spending much of their time underwater, invisible to the human eye. Most of the early 'scientific' descriptions of whales came from stranded animals, many of which were immortalised in drawings and engravings, although the animals' details were often exaggerated and erroneous. In his 1851 novel *Moby-Dick*, Herman Melville addressed this issue when he wrote, 'Consider! Most of the scientific drawings of whales have been taken from the stranded fish; and these are about as correct as a drawing of a wrecked ship, with broken back, would correctly represent the noble animal itself'.

Perhaps the most accurate early descriptions of whales came from the writings of the Greek naturalist and philosopher Aristotle. In his *Historia Animalium* (*c.* 350 BC), he recognised the cetaceans' mammalian characteristics.

The dolphin, the whale, and all the rest of the Cetacea, all, that is to say, that are provided with a blow-hole instead of gills, are viviparous. That is to say, no one of all these fishes is ever seen to be supplied with eggs, but directly with an embryo from whose differentiation comes the fish, just as in the case of mankind and the viviparous quadrupeds.

D'Arcy Wentworth Thompson (transl.), *History of Animals* (1910)

The North Sea coast of Holland was a particular hotspot for whale strandings, with at least 40 strandings of various species recorded between the 1530s and 1690s. This engraving by Frederik Ruysch depicts a sperm whale that beached in 1577.

Aristotle also made observations on dolphin biology and behaviour: 'The dolphin and the porpoise are provided with milk, and suckle their young … Its period of gestation is ten months. It brings forth its young [in] summer, and never at any other season … Its young accompany it for a considerable period; and, in fact, the creature is remarkable for the strength of its parental affection.' In addition, he noted the dolphin's lifespan: 'It lives for many years; some are known to have lived for more than twenty-five, and some for thirty years.' While not perfect, Aristotle's descriptions were surprisingly correct and provided some of the most accurate information on cetaceans for centuries.

Bits and pieces of information about whales continued to be recorded. In 13th-century Norway, for example, an account known as *The King's Mirror* listed sea creatures seen off the coast of Iceland. Aside from various oddities, it described species that are recognisable today, including the narwhal, orca, sperm and right whales, some of which it says were evil and longed to eat humans, while others were 'good' and would protect people who did them no harm. The right whale's description, like most of its time, was a mix of accuracies combined with rather outlandish falsehoods:

> people say that it subsists wholly on mist and rain and whatever falls upon the sea from the air above. When one is caught and its entrails are examined, nothing is found in its abdomen like what is found in other fishes that take food, for the abdomen is empty and clean. It cannot readily open and close its mouth, for the whalebone which grows in it will rise and stand upright in the mouth when it is opened wide; and consequently whales of this type perish because of their inability to close the mouth. This whale rarely gives trouble to ships. It has no teeth and is a fat fish and good to eat.

> I. Whitaker (ed.), *The King's Mirror* (1985)

Ironically, however, it was whaling that opened doors to new information about cetaceans. Aboriginal whalers certainly needed to have an understanding of the biology, behaviour and migratory patterns of the species they were hunting, and as whaling spread and became a commercial enterprise, information was expanded to include new species that were being targeted.

In the late 17th century, Antonie van Leeuwenhoek made improvements to the microscope and used it to examine parts of whales brought to him by friends and people associated with the Dutch whaling industry. He noted the different layers of flesh and also that whale testicles were as large as a 'Firkin of Butter that weighs about a hundred Weight'.

One of the most influential early books on whales and whaling, *An Account of the Arctic Regions with a History and Description of the Northern Whale-fishery*, was written by William Scoresby, the son of a whaling captain. After working on board his father's ship, he studied anatomy and chemistry at Edinburgh University and then went on to captain his own whaling ship, the *Resolution*, from 1810 to 1823. Scoresby's book, published in 1820, focused on the 'Greenland' (bowhead) whale industry and, along with giving descriptions of the whales and Arctic whaling, lamented the decimation of the bowhead populations.

In 1864, John Edward Gray published *On the Cetacea Which Have Been Observed in the Sea Surrounding the British Isles* and followed that with a number of works in which he listed and described cetaceans. Gray was amongst those who noted the difficulties of the work: 'There is no series of large animals more difficult to observe and to describe than the Whales and Dolphins … They are only seen alive at a distance from the observer, and generally in rapid motion and under unfavourable circumstances for study.' Despite these difficulties, scientists and naturalists remained fascinated with whales and continued their research, often under trying circumstances.

In the southern oceans, one of the earliest scientific studies on whales was carried out during the Belgian Antarctic Expedition of 1897–99, when Emile Racoviţă collated Antarctic whale sightings in order to provide information on their distribution. During the era of commercial whaling, however, most of the information about large whales came from scientists working on the slippery decks of whaling ships or the bloody floors of whaling factories. American naturalist Roy Chapman Andrews noted that until the establishment of shore whaling stations, few scientists had had the opportunity to examine more than five or six whales during their lifetime.

FASCINATING FACT

In 1921, an American scientist examining a humpback whale specimen discovered hind limbs extending 1.3 metres outside of the whale's body. The tiny 'legs' had a tibia, tarsus and metatarsal, the same bones as those found in a land mammal's lower legs and feet.

The examination of whale carcasses gave biologists information about whale physiology, anatomy and reproduction but told them little about whale behaviour and social interactions, although many of the researchers on whaling ships did note the migration routes of different species. Once again, the irony is that many of the early studies were undertaken in order to aid whaling.

In 1817, the British Colonial Office established the so-called Discovery Committee, one of whose main goals was to conduct research that would facilitate the preservation of the whaling industry. The committee was suspended during the First World War, but when it reconvened in the mid-1920s it initiated tagging programmes of live whales, which gave both scientists and whalers additional information on whale movements. This involved firing a stainless-steel marker known as a discovery tag into the body of a live whale. Each tag was individually numbered, and scientists kept careful records of when and where a whale was marked. Between 1925 and 1934, the Discovery Committee's research ship *William Scoresby* 'marked' up to 3000 whales, mainly in Southern Ocean waters south of Australia. The tagging method itself was invasive to be sure but not nearly as invasive as the tag's recovery, which occurred only when the whale was harpooned and killed.

This photo of the research ship William Scoresby *appeared in the* Auckland Weekly News *on 10 November 1937 just prior to the ship's departure to Antarctic waters to study the movement of whales.*

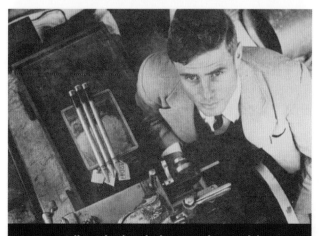

Scientist Bill Dawbin beside three stainless steel discovery tags. One of the early indications that Soviet whalers might be falsifying data occurred in 1965 when Australian biologist Graeme Chittleborough reported his findings that two numbered discovery tags, supposedly recovered from a sperm and fin whale by Russian whalers, were in fact tags that had been fired into humpback whales.

CETACEAN IDENTIFICATION

As whaling ceased and the 'Save the whales' movement grew, there was an increased hunger to learn more about our fellow mammals. However, scientists now had few dead whales to examine and the live ones posed a number of problems, among them the fact that they all looked alike and they all kept disappearing from view. A huge breakthrough occurred in the 1970s when researchers discovered that killer, humpback and right whales could be individually identified through variations in their natural markings.

In the early 1970s, while conducting a census of killer whales in Canada's Pacific Northwest, marine biologist Michael Bigg realised that each orca had a unique greyish-white saddle patch behind its dorsal fin. This discovery ultimately enabled him to identify every individual orca in his study area and led to the establishment of a similar photographic identification study of the resident population of orca living in nearby waters in the USA's Pacific Northwest. Both the northern and southern resident orca populations have been photographically identified as part of a census count every year since the beginning of the 1970s programme, and the use of photo identification has extended to studies of orca and other cetacean species around the world.

Thousands of miles away, and also in the early 1970s, researchers studying North Atlantic humpbacks in the Gulf of Maine realised that each whale had black and white pigmentation on the underside of its tail that was unique to each individual. Using photo identification, more than 6000 whales have now been catalogued in the North Atlantic Humpback Whale Catalogues and family trees established for many. By following these whales over the decades, researchers have learned about the age that females start to calve and their calving intervals, and whether the calves return to the same breeding and feeding grounds. In addition, researchers have gained new information on the movements and migration routes of the North Atlantic population.

During the 1970s, research was also taking place on southern right whales in Argentina. It wasn't long before the scientists involved realised that each whale could be identified individually from the unique arrangement of callosities found on its head. Callosities are somewhat like warts on a human, being patches of rough raised skin that are present when the whale is born, and each callosity essentially remains in the same position on the whale's head throughout its adult life.

Photo identification is now used with a number of different cetaceans by researchers around the world, although the form of identification varies from species to species: blue whales display unique bluish-grey mottling patterns on their body; pilot whales have a distinctive white 'cape' behind their dorsal fin; many beaked whales have distinguishing scars from cookiecutter shark bites; and fin whales, along with many other whale and dolphin species, can be identified from the size and shape of their dorsal fin. In addition, researchers are able to identify individuals from photographs of scars, nicks or other distinctive characteristics. By using photo identification, scientists have been able to obtain information on a population's migration and movement patterns, social affiliations and family histories, age at maturity and calving intervals, birth and death rates, and an accurate count of numbers.

Orca with open saddle patches are usually easier to tell apart. As a result of years of photographic identification, researchers not only know the overall population of orca in their areas, they also know the number of pods, their composition and their social organisation, along with each whale's genealogy.

FASCINATING FACT

During the first years of his study, Marine biologist Michael Bigg realised there are two distinct types of killer whales in the Pacific Northwest: 'residents', which feed mainly on fish; and 'transients', which feed mainly on other marine mammals. A third type, 'offshores', has since been identified. Currently, there is a push to recognise transient orca populations as a separate species, to be known as Bigg's whales.

Every individually identified whale is given an identification number and sometimes a name, which helps researchers remember it. These southern right whales were identified in the Auckland Islands in 1999, 'Smooth Dude' is on the left, and 'Castle' on the right.

When a humpback whale lifts its tail prior to a dive, it displays its individual pigmentation pattern. Many humpbacks also have distinctive scalloping along the trailing edge of the tail.

To obtain a sample of skin for DNA analysis, researchers fire a dart into the body of the whale, which then bounces off with a tiny (around 5-millimetre wide) plug of skin attached to its tip. Researchers can also use abrasive material, such as an ordinary pot scrubber, attached to a pole to rub off a piece of the whale's skin.

Corkscrew, a male orca from Aotearoa, was photographed and named by the author in 1984. Since then, researchers have photographically identified the majority of orca that reside in New Zealand waters.

Some whale populations have been photographically identified over periods of 30–40 years, allowing researchers to document and record amazing life histories about their subjects.

FASCINATING FACT

Salt, one of the first individual whales identified in the Gulf of Maine, has been photographed every year since 1976. Salt is now considered to be the Grande Dame of the North Atlantic population. As of 2010, she had given birth to twelve calves and was a grandmother of eight.

DNA ANALYSIS

New advances in molecular biology have made DNA analysis one of the most important tools used in the study of whales today. A tiny sample of whale skin (obtained by firing a small dart at the animal) or sample of faeces is all that is required, but this is equivalent to the Holy Grail for whale researchers and can provide answers to questions that are not obtainable by any other means. Not only can the species of whale be identified but also the individual, its sex and age, its relatedness to other whales and its relationship to other whale populations. Since DNA doesn't change over time, each skin sample becomes a genetic identity tag for that particular whale or dolphin that lasts throughout its entire lifetime. Further laboratory analysis of a whale's skin tissue can also give information about its diet and about pollutants that are present in its environment.

LISTENING TO WHALES

Acoustic research has revealed some fascinating information about many whale species, although there is still much more to learn. For example, scientists have discovered that humpbacks sing songs that vary between populations and change over time, although the exact purpose of the songs and why they change is still being debated. They have also learnt that sperm whales use coda clicks to communicate, although again what they are communicating is still unknown. Acoustic studies in the Pacific Northwest have revealed that each killer whale pod has its own distinct dialect, allowing researchers to identify which pod is in or approaching an area just by listening to their calls on speakers attached to hydrophones (underwater microphones).

Acoustic research has also opened new doors when it comes to locating whales. In 1992, the US Navy allowed whale researchers to access its submarine-tracking stations. The stations' sophisticated listening devices can pick up very low-frequency sounds made by blue and fin whales, as well as calls from other whale species. Using the equipment, researchers have been able to track whale movements over entire ocean basins and sometimes follow individual whales over many hundreds or even thousands of kilometres. In 2013, acoustic technology was used for the first time in the Southern Ocean when the Antarctic Blue Whale Project utilised passive acoustic sonobuoys to successfully locate, track and study blue whales in a sample area west of the Ross Sea. During the seven-week voyage, the researchers made more than 600 hours of recordings and analysed more than 26,500 blue whale calls. The directional sonobuoys allowed the acousticians to triangulate a whale's position from its vocalisations and then direct the research ship to its location. The international research team collected 57 photo identifications, 23 biopsy samples, and attached satellite tags to two blue whales. The acoustic research was so successful that it is likely to become a blueprint for use in the study of far-ranging whale populations throughout the world.

Whale or dolphin movements can also be tracked from land directly through the use of a theodolite (surveying tool) or indirectly from a tiny satellite tag attached to the whale's body. The tag transmits a signal to a satellite whenever the whale surfaces, and scientists monitoring the signals are able to track the whale even when it is far from land in inaccessible ocean environments. The tag not only discloses the whale's movements but also gives information on diving behaviour and respiration rates. Projects such as the Greenpeace-funded Great Whale Trail, in which South Pacific humpbacks were tracked via satellite tagging, have revealed answers to questions relating to the movements and behaviour of many species.

FASCINATING FACT

In the first nine months of using the US Navy's listening devices, researchers had recorded tens of thousands of blue, fin, humpback and minke whale calls.

RESEARCH IN AOTEAROA NEW ZEALAND

As in other parts of the world, New Zealand's early cetacean research took place during the era of whaling. Between the years 1949 and 1961, the majority of the work was conducted on humpback whales. During this time, whalers at the Perano station in the Marlborough Sounds worked closely with biologists such as William Dawbin, who examined dead whales on the factory floor and tagged live whales from the decks of their whaling vessels. While most of the migrating whales were believed to be heading to breeding grounds around South Pacific islands, a few humpbacks that were tagged in Tory Channel ended up being harpooned in eastern Australia, providing the first proof that some individuals passed through New Zealand waters and then made their way to Australian coasts.

By 1962, the humpbacks were 'finished' in New Zealand waters and the scientists turned their attention to sperm whales. Anti-whaling sentiments had not yet been born and there were no immediate thoughts about ending whaling; the main purpose of the sperm whale research was to avoid the crash in numbers that had occurred in the humpback and right whale populations. In his paper 'New Zealand Whaling and Whale Research, 1962–1964', published in *New Zealand Science Review* in 1965, biologist D.E. Gaskin stated, 'if the population [of sperm whales] is to be exploited indefinitely, it is necessary to obtain data which will throw light on the structure and dynamics of the population so that exploitation rates do not exceed replenishment.'

Simple underwater listening devices, such as this hydrophone, can be used to listen to and record whale and dolphin vocalisations underwater.

From 1963 to 1964, intensive surveys were carried out on sperm whale movements and at their concentrations on whale grounds around mainland New Zealand and areas to the north (the Kermadecs and Fiji), south (Campbell and Auckland islands) and east (Chatham Islands). In addition, sperm whales killed by whalers were examined and detailed reports were published about their size, sex and stomach contents. During the surveys, other large whales and smaller dolphin species were also sighted and logged, and a number of large whales were tagged. Sperm whaling in New Zealand ended in 1964, not because of a lack of numbers but because a worldwide drop in the price of whale oil meant it was no longer economically viable.

The sperm whale surveys had some positive side effects, as Gaskin pointed out: 'the first right whale sightings in over 30 years were documented and a considerable body of data was collected on the species of dolphin around New Zealand.' Ironically, the intensive studies of sperm whales to determine how many could be sustainably killed laid the foundation for future studies on other cetacean species found in New Zealand waters. The Gaskin paper went on to say:

> Between the great whales which are hunted commercially and the small dolphins are nearly a dozen whale species of intermediate size with no

commercial value in this country. In most cases, little is known about the sizes of their populations, their movements or their habits, yet many of these species are equally or more interesting than their large relatives.

Of particular interest are the strange strap-toothed whale with projecting tusks, the much feared killer whale with its huge dorsal fin, the shark like pygmy sperm whale as well as the common 'blackfish' or pilot whale which sometimes strands itself on the coasts in such numbers that the local body responsible for the disposal of perhaps two hundred dead carcasses will long remember the event.

By 1965, whaling in New Zealand was over, but the interest in discovering new information about different whale species was just in its infancy. The dedication and efforts of a small handful of pioneering whale scientists during the whaling years laid the groundwork for the researchers who were to follow.

HUMPBACK WHALE RESEARCH

During the Perano whaling years (1911–64) in Cook Strait, the humpback season operated for about 13 weeks, from

Early researchers, including naturalist Sir James Hector seated on the chair, with a skeleton from a whale captured at Stewart Island in January 1874.

mid-May to late August, when the whales were making their northern migration to their mating grounds. The whalers spotted their quarry from the top of a small peninsula overlooking Cook Strait. For the first few years, the men hunkered down on the frigid, windy hilltop in a small dugout area, but gradually a hut was constructed to afford them some shelter from the elements. From dawn until dusk, the whalers scanned the horizon in search of whales. When one was sighted, a cry of 'thar she blows' rang out and a flag was raised to alert other whalers waiting below, who then raced to their boats and set off in pursuit of their quarry.

Since the crash in humpback numbers in the early 1960s and their subsequent international protection from commercial whaling in 1966, very few humpback whales have been sighted in New Zealand waters. In 2004, the Department of Conservation (DOC) began the annual Cook Strait Humpback Whale Survey to determine how many whales are still migrating through the area. The survey ran for two weeks during the first four years, and in 2008 it was extended to four weeks. The research team enlisted the aid of former whalers and a lookout was set up on DOC land on a high hill adjoining Okukari Farm, belonging to fifth-generation whaler Joe Heberley. Just as in the old days, the spotters shout 'thar she blows' and the boatmen waiting below (now researchers) set off in pursuit of their quarry. Like the whalers of old, the team hope to shoot the whales, but their weapons – cameras and DNA darts – are more benign.

The Cook Strait Humpback Research team photographs as many whales as possible during their survey. The photos are then compared to those taken by other Oceania humpback whale scientists in an attempt to discover which breeding ground New Zealand's migrating whales are travelling to. The photos also show if the same individuals are migrating through New Zealand waters every year.

Between 2004 and 2013, a total of 439 humpback whales were sighted by survey researchers, of which 120 whales were photographically identified and 190 DNA samples obtained. The best season was 2012, when 106 humpbacks were spotted. Photo identification and DNA samples have so far matched a few of the humpbacks passing through New Zealand waters with whales in New Caledonia and eastern Australia (Byron Bay, Hervey Bay and Ballina). While the 106 humpbacks counted in 2012 were a record for the survey, the number was still pitifully small compared to researchers' counts of 70–75 whales per day migrating along Australia's east coast. New Zealand's migrating humpback population is increasing, but it still has a long way to go before it completely recovers from the whaling years.

Whalers in a lookout station, 1950s.

The first and only cow–calf pair recorded to date in the Cook Strait Humpback Survey was spotted in 2010 by Johnny Norton. As of 2013, only three individuals had been resighted over the six-year study.

SOUTHERN RIGHT WHALE RESEARCH

The humpback is not the only whale species in Aotearoa whose numbers are still affected by the whaling era – southern right whales also continue to be a rare sight in mainland waters. An estimated 16,000 right whales may have once mated and calved in New Zealand waters; at its lowest point, around 1913, the population was reduced to 14–52 individuals. The first sighting of a right whale in over 35 years ironically occurred at the Tory Channel whaling station in 1963. That same year there was a grand total of five mainland sightings of the whale that was once said to be as common as mud.

Between 1963 and 1993, right whale sightings around mainland New Zealand did not increase dramatically and research into the species was virtually non-existent – it was considered that there were too few whales to warrant study.

Aside from photographing the humpbacks, researchers try to obtain a skin sample from each whale by shooting a dart into its body with a specially designed rifle. The sample is then fixed in alcohol and sealed in an airtight container for transportation to a lab for DNA analysis.

In his 1964 paper in *Tuatara*, titled 'Return of the southern right whale (*Eubalaena australis Desm.*) to New Zealand waters 1963', D.E. Gaskin reported:

> During the 1963 season, the catastrophic decline in the number of humpback whales passing the New Zealand coast was again noted … the blue whale has also declined in numbers until it is now in severe danger. The damage to these two species has been done despite controls laid down by the IWC. Even the stocks of finbacks are declining year after year. A similar decline in these species has been noted in Australian waters. Despite the declining numbers of these three species, the rarest large whale in the Southern Hemisphere is still the Right Whale… it is doubtful whether the number of individuals exceeds a very few hundred in the whole of the eastern Southern Ocean even though it has been completely protected there for over 25 years.

There were, however, reports indicating that 'significant numbers' of right whales were present around the subantarctic islands. Acting on this, a team of researchers set up camp in the winter of 1995 at Northwest Bay on Campbell Island with the purpose of assessing the remnant population of right whales still surviving in the waters of the subantarctic islands. That same year, another research team travelled to the Auckland Islands, also in search

Researchers at Campbell Island camped on land and, when weather conditions permitted, took to the seas in a 3.8-metre boat in order to photographically identify the 15–17-metre whales that were present in their Northwest Bay study area.

of right whales. As in the later Cook Strait Humpback Survey, the right whale researchers hunted their quarry with cameras and DNA darts. Project Tohorā's Campbell Island team concentrated on photo identification, while Auckland University's Auckland Island team collected skin samples for DNA analysis as well as photographs. Both research teams conducted census counts, and kept detailed records of behaviours and group compositions.

Winter right whale research in the subantarctic continued every year until 1998. At Campbell Island, the majority of whales appeared to be adults and, while courtship groups were frequent, no calves were sighted. At Auckland Island, a mix of ages was observed, ranging from new calves to sub-adults and mature adults. The maximum number of whales sighted at Campbell Island during a winter season was 44 individuals (1995), while the numbers at Auckland Island were significantly higher, with more than 125 whales sighted in 1997. After four years of study, it became apparent that although the whales were utilising both islands, the vast majority of mating and calving was occurring at Port Ross in the Auckland Islands.

On the New Zealand mainland, a total of just four right whales were sighted in 1979 and 1980. In 1991, one cow made the headlines when it was observed with two calves

This female southern right whale was sighted off Kaikoura in 1991. The sighting was particularly interesting as the female was accompanied by two calves – one very small and another that was larger. Twins are uncommon; perhaps the mother had taken on an orphaned calf, but the reason why there were two youngsters with her is not known.

Researchers at Campbell Island, New Zealand's southernmost subantarctic island, joked that working there was comparable to doing 'research in an icebox'.

off Kaikoura, and from 1991 to 2002, 11 cow–calf pairs were sighted, with a grand total of 22 whales reported in 2000 alone.

From 2003 to 2010, 125 sightings were reported from the mainland, which included 28 cow–calf pairs. A tagged right whale was tracked by satellite from the subantarctic to the South Island in the winter of 2009, and researchers announced in 2011 that, for the first time, DNA matches had been made between seven subantarctic and mainland right whales. In 2013, a further eight photograph, and three DNA, matches were made between the two areas. While the matching of these relatively few individuals may seem insignificant, it does indicate that the subantarctic population may be starting to recolonise its former mainland New Zealand habitat. The first evidence of calving site fidelity in recent years has now been documented, with

two identified females sighted with a new calf around the mainland at four-year intervals, the first in 2005 and 2009, and the second in 2006 and 2010. As well, a few groups of mixed sex whales, potentially mating groups, have been reported. While the increased numbers are encouraging, they are still relatively pathetic when compared with the estimated pre-whaling population of 16,000 and it will be many years before the mainland southern right whale will once again be considered common.

SPERM WHALE RESEARCH

While New Zealand sperm whale numbers never fell as low as those of the humpback and southern right whales, there is no doubt that the populations were diminished by whalers, who killed more than 390,000 in the southern hemisphere between 1904 and 1983. In the late 1870s,

THE RIGHT IDENTIFICATION

The raised areas of rough skin on a right whale's head are known as callosities. These grey patches are inhabited by a large variety of organisms, including colonies of parasitic crustaceans known as cyamids, turning them white, yellow and sometimes a pinky orange, which makes them stand out against the whale's black skin. Each whale has its own distinctive pattern of callosities, which can therefore be used to identify individuals. The right whale research teams on the subantarctic islands identified over 300 individual whales during their 1995–98 study and also obtained matches between the Campbell and Auckland island groups, proving that the whales were from the same

population. The Auckland Islands' catalogue of identified whales from the 2006–11 surveys now contains at least 513 individuals. Recent genetic evidence strongly suggests that the subantarctic and mainland population of southern right whales represent one stock, with the population using the subantarctic islands as their primary mating and calving wintering grounds while slowly recolonising the mainland as a secondary wintering ground. Based on the current scientific analysis, New Zealand's southern right whale population is estimated at 1300–2400 individuals. Studies are continuing to determine more precise population estimates.

'Spunky', ID # 62, photographed July 1996 at Auckland Islands and August 1997 at Campbell Island.

'Mr Cool', ID # 18, photographed 25 July 1995 at Campbell Island and 8 August 1995 at Auckland Islands.

Elephant Ears, so named because of his huge, almost floppy tail, was one of the first individual sperm whales identified off Kaikoura. He was resighted every year from 1983 to 1999, making him one of the longest known returning residents to the region.

off the Chatham Islands, the whaling ship *Splendid* from Dunedin saw 'a school of sperm whales, ten miles square, took nine and saved 800 barrels of sperm oil'. Sadly, there is little chance today, after the years of industrial whaling, that such a sight would be encountered anywhere in New Zealand waters.

Research on sperm whales was on-going during the 1963–64 sperm whaling season, when 241 male and seven female sperm whales that had been harpooned were examined and 27–30 live whales were tagged. Papers were subsequently published by D.E. Gaskin on the behaviour, distribution and social structure of sperm whales, and by Gaskin and M.W. Cawthorn on the species' diet and feeding habits.

The whales' stomach contents indicated they had been feeding mainly on squid but grouper and ling were also present in significant numbers at certain times of the year, along with other fish species such as kingfish, and various species of shark, dory and eel. One whale's stomach contents included a rather 'mangled unidentified octopus', while whole coconuts and pieces of coal, pumice and wood were present in other stomachs.

In his research papers, Gaskin noted the presence of young bachelor males around the mainland: 'some

of the males in the Cook Strait region may belong to a non-breeding sub population... over 70% of the whales sighted from the whale chasers are bachelor males with an average length of 13.4m'. Today, the majority of sperm whales sighted in mainland New Zealand waters are still bachelor males, many of which are observed in deep-water feeding grounds off Kaikoura on the South Island's east coast. The relatively easy accessibility of these feeding grounds has attracted numerous researchers from New Zealand and around the world to study the sperm whales, of which more than 150 individuals have been identified. Unlike a humpback's tail, the underside of a sperm whale tail usually lacks visible marking and is solid black. This initially posed problems for scientists trying to identify the whales, until they realised the scalloping pattern along the edge of the tail is unique to each individual.

Research into sperm whales off Kaikoura has continued, with scientists not only collecting photographs of individuals but also using hydrophone (underwater microphone) arrays linked to computers to record the whales' echolocation clicks. The researchers have been working on a way to calculate the size of an individual whale based on its click rate, and also on ways to determine the number of whales feeding in

Sperm whaling ship the Splendid *pictured between 1870 and 1890. In the late 1870s the ship caught nine sperm whales in one outing.*

the area from an analysis of their echolocation clicks. Underwater movements of whales can also be tracked using the hydrophone arrays. In 2013, USA and New Zealand scientists attached small recorders to the backs of Kaikoura's sperm whales using suction cups. The researchers then were able to follow the whales on their computers as they dove into the deep Kaikoura canyons in search of their prey. This type of new technology gives us a picture of the whale's life that has never before been obtainable. Results from the 2013 research project will be published in the not too distant future.

RESEARCH INTO OTHER CETACEANS

Scientists are conducting research on many of the other whale and dolphin species found in New Zealand waters, including Bryde's, killer and pilot whales, and Hector's, common, dusky and bottlenose dolphins. Scientists also opportunistically collect information and skin samples from cetaceans that strand on New Zealand coastlines, many of which include beaked whale species that are rarely sighted in the wild. Many of the researchers are affiliated with universities such as Otago, Auckland and Massey, or work through overseas universities,

Ingrid Visser from the Orca Research Trust photographs the saddle patch of a male orca off the Kaikoura coast. This male is a member of a pod that has been photographed around both the North and South islands.

while others work with research organisations such as the New Zealand Whale and Dolphin Trust, the Orca Research Trust, WWF-New Zealand, Kaikoura Ocean Research Institute, the Centre for Cetacean Research and Conservation in the Cook Islands and the Department of Conservation, to name just a few.

ALAN BAKER:

A life of cetacean research

I grew up in the 1950s in the Bay of Islands, where I spent many happy hours fishing from my father's launch and enjoying the marine environment. We regularly saw humpback and Bryde's whales, common and bottlenose dolphins, orca and even the occasional southern right whale. Learning to scuba dive in the late 1950s opened up a whole new vista for me, and I made plans to begin a career in marine science.

I studied at Victoria University in Wellington in the 1960s, where, after a three-month expedition to Antarctica, I concentrated on pelagic and deep-sea animals. During my post-graduate research on pilchards, I was introduced for the first time to endemic Hector's dolphins, which came to the research vessel at night to pick pilchards from my net. My world was full of cetaceans: orca, bottlenose and dusky dolphins often appeared near the vessel, and I could also view them from my window at the university's marine laboratory in Island Bay.

In 1969, I joined the Dominion Museum as a marine scientist and soon realised that the museum (now known as Te Papa) held many fascinating

and important cetacean specimens. New Zealand waters contain some of the rarest cetaceans in the world, and my attention again focused on the tiny dolphin that had once stolen pilchards from my net. At the time, the information about Hector's dolphin (pictured opposite) was vague and speculative. My first paper on the species, written jointly in 1973 with W.F.J. Mörzer-Bruyns, a Dutch sea captain who had cruised New Zealand's coasts for many years observing cetaceans, set the scene for the very first field study of the dolphins, which began in Cloudy Bay in 1978. Many projects were to follow, and during my 25 years at the museum (of which I eventually become director), I produced guidebooks on Australasian whales and dolphins, and researched and wrote papers on the cetaceans of the New Zealand region, including the subantarctic and Antarctic oceans.

In 1993, I left the museum and joined the Department of Conservation (DOC), where I contributed to several papers on beaked whales as well as other marine mammal species. I began a four-year aerial field study of Bryde's whales in the coastal waters of northeastern New Zealand in 1999. During this time, I was asked by DOC to be involved once again with Hector's dolphin. In 2002, after completing a number of tests, statistician Adam Smith, geneticist Franz Pichler

new projects with foreign scientists and to contribute to several papers, including a major publication on Odontocetes of the Southern Ocean Whale Sanctuary and another on vessel collisions with cetaceans. In 2005, I travelled to Namibia and observed the South African cousin of Hector's dolphin — Heaviside's dolphin. Given that I had already been fortunate enough to view Hector's South American cousins, Commerson's and Chilean dolphins, in Tierra del Fuego, I was pleased to add this fourth *Cephalorhynchus* species to my list.

After contributing to a paper on blue whales in the southern hemisphere in 2007 and writing a chapter on cetaceans in 2008 for the New Zealand Inventory of Biodiversity, I really have now retired and gone fishing!

and I confirmed that the North Island Hector's population was morphologically and genetically distinct from the South Island population and therefore fitted the criteria to be considered a separate subspecies. We named the subspecies after the mythical Māori hero Maui, who in the creation of Aotearoa, fished up the North Island, Te Ika a Maui. The highly endangered North Island subspecies is now officially known as *Cephalorhynchus hectori maui*, or Maui's dolphin. Also in 2002, I was involved with an international research group that described a new beaked whale species, *Mesoplodon perrini*, or Perrin's beaked whale.

In 2003, I retired from DOC and moved back to the Bay of Islands to go fishing, only to come immediately out of retirement to work on several

RESEARCH IN THE SOUTH PACIFIC

Scientists in New Zealand collaborate with others from around the world, particularly in the South Pacific, where the South Pacific Whale Research Consortium (SPWRC) was established in 2001 to conduct non-lethal research on humpback whales and other cetaceans living in the region. By 2010, SPWRC scientists had photographically identified over 2000 individual whales at overwintering grounds in the South Pacific. The researchers have found that almost half of Oceania's breeding population of humpbacks frequent the islands of Tonga, while the next largest groups mate and calve in New Caledonia and French Polynesia. In 2012, SPWRC researchers estimated the total Oceania humpback population to be about 4300 whales. Despite covering an enormous area, the islands of the South Pacific remain the least populous whale breeding grounds in the southern hemisphere, the population having a recovery rate of about 5 per cent per annum. The Oceania humpback still lags behind the eastern Australian humpback population, which has been recovering at a rate of 10–11 per cent a year and was estimated at around 15,000 individuals in 2012.

In addition to their research on whale numbers and behaviour, SPWRC scientists have been recording humpback whale songs. While it has long been known that humpback males 'sing' on their breeding grounds, an exciting new discovery about the 'cultural' transmission of their songs throughout Oceania has only recently been unveiled.

Mike Noad and Ellen Garland from the University of Queensland and several of their colleagues have studied a collection of 775 humpback whale songs recorded by the SPWRC scientists, publishing their astonishing findings in 2011. Their analysis revealed that between the years 1998–2008, 11 different song types were sung by the males. The surprising discovery was that eight of those song types had originated in Australia and then travelled in an easterly direction across the South Pacific, with four of the songs spreading across the entire region. The songs appeared to move initially from eastern Australia to New Caledonia and Tonga, and then on to American Samoa and the Cook Islands, with four songs ultimately travelling as far as French Polynesia.

The discovery was surprising because there is little evidence of an interchange between the humpbacks in Australia and those in Oceania. Most humpback whales

Researchers Ygor Geyer and Nan Hauser use a windsurfing mast to attach a satellite tag to and take a DNA sample from a humpback whale travelling west towards the Tongan Trench.

exhibit a high level of site fidelity, which simply means that they return year after year to the same mating and calving area. Yet somehow, four long-distance song types that originated in eastern Australia had made their way more than 6000 kilometres across the Pacific, where they were ultimately being sung by humpback males on French Polynesian breeding grounds. So how were the songs being transmitted across such vast distances?

One possible scenario is that a few roving males from Australia were making their way eastward into Oceania's waters, where their song was taken up by the South Pacific males and passed on like Chinese whispers. It has already been demonstrated on other mating grounds that the presence of just one or two new males in a population will trigger a song change. Another possibility is that the songs were passed on during the humpbacks' journey to the tropics. Although most singing is done on the breeding grounds, males have been recorded singing as they migrate, and their song may have been heard and picked up by other males that were sharing migratory routes during a portion of their northward journey.

The researchers noted in their paper that 'This is the first documentation of a repeated, dynamic cultural change occurring across multiple populations at such a large geographic scale.' Of interest also is the fact that the songs are travelling in an eastward direction, echoing the pathway taken by early human migrations across the Pacific.

The information gained from scientific research into cetaceans is a powerful tool, giving us increased awareness of, and information about, the animals themselves, along with knowledge of specific threats to their existence. In 1964, for example, D.E. Gaskin referred to the 'much feared killer whale'. Today, the vast information gained from research into this species has completely changed the way we view it; we no longer think of the largest species of dolphin as fierce killers but rather as extremely intelligent and beautiful animals that are supreme predators in our oceans.

Research into Hector's dolphins in the 1980s suggested that 230 had died in set nets around Banks Peninsula between 1984 and 1988. The estimated population of the dolphins in the area was only 740 – clearly too many dolphins were dying. The knowledge gained from the scientific study increased public awareness, which led to empathy and a desire to make a change. The result was the creation of the Banks Peninsula Marine Sanctuary, which prohibits or limits set netting. While all this sounds simple, the reality is that it takes years of careful study to gather data; the changes that need to happen do not take place overnight.

Today, modern research techniques such as photo identification, DNA analysis, radio and satellite tagging, and sophisticated listening devices that can track whales over vast ocean distances have opened new windows into the world of whales. Scientists from around the globe are using these innovations to discover exciting new information about cetaceans, as well as to identify the specific threats these marine mammals are facing both at present and into the future.

Dr Marc Oremus extracting DNA from dolphin tissue samples at the University of Auckland.

CHANGING TUNE

During the 1996 breeding season, researchers off the eastern coast of Australia recorded 82 individual humpback whales all singing the same song. But, the researchers were confused when they also recorded two males singing a completely different song. As the season progressed, all of the whales started to adopt the new tune, and when the 1997 breeding season rolled around the returning east coast whales had switched over completely to the new song. It turned out that the song adopted by the east coast whales was the same one being sung thousands of kilometres away by humpback on Australia's west coast. The question is, why did the two males decide to leave the Indian Ocean to visit the Pacific Ocean breeding grounds, and why did the Pacific Ocean whales drop their old song completely in favour of the Indian Ocean melody?

WHALE CONSERVATION

Whales, like Pacific Island people, are ocean voyagers and their tremendous migrations reflect many of our own journeys and ocean crossings. Whales are indeed part of our rich Pacific heritage. Whales are indeed part of us. Yet today, it is fair to describe great whales, including the humpbacks, as a mere remnant of what we once shared our ocean with.

The story of what has happened in a mere few hundred years and the devastation wrought by whaling, the impacts of which we are still living with today and will be inherited by our children and the generations that are to come, is a story that must be told, not just for whales, but for ocean survival, and indeed our survival.

HRH Princess Salote Mafile'o Pilolevu Tuita of Tonga (April 2010)

It took millions of years for whales to evolve to their present form but only a few hundred years to bring some species to the brink of extinction. By the 1850s, right whales had been virtually exterminated in the waters around Aotearoa and Australia, and indeed throughout the world, and by the 1970s a breeding population of humpbacks in the South Pacific was almost non-existent. Sperm, fin, sei and blue whales also suffered at the hands and harpoons of whalers, yet despite warnings of their demise, whaling countries continued to kill whales until, in some cases, there were simply not enough left to make hunting them economically viable.

THE INTERNATIONAL WHALING COMMISSION

New Zealand was one of the founding members of the International Whaling Commission (IWC), established in 1946. The commission's original mandate under the International Convention for the Regulation of Whaling (ICRW) was not to protect whales but to regulate 'whale stocks and thus making possible the orderly development of the whaling industry'. In essence, its job at that time was to set quotas for different whale stocks that would allow for the sustainable 'harvesting' of various species.

As whale populations declined and consciousness towards saving the mammals has grown, however, the IWC has evolved to become the main international forum for debate as to whether or not whaling should be allowed to continue at all. Since the late 1960s, the floor of the IWC has been a battleground between conservationists who wish to protect whales and representatives of those nations that wish to continue whaling. Although New Zealand left the commission in 1968 following its own cessation of whaling, it rejoined in 1976 when the IWC changed its focus to conservation. Today, New Zealand remains a staunch advocate of

Minke whales lined up in a whaling ship in Antarctic waters.

whale conservation and its representatives attend every IWC meeting to argue against whaling and to support marine sanctuaries in the southern oceans. In 1982, victory for anti-whaling advocates seemed imminent when the IWC declared that a moratorium on commercial whaling would go into effect at the beginning of 1986. Loopholes in the legislation, however (particularly Article 8 of the ICRW), have enabled Japan to continue to hunt and kill whales for the purpose of 'scientific research', and Norway and Iceland to continue commercial whaling.

Another victory appeared to be won by the anti-whaling lobby in 1994, when members of the IWC voted to establish a 50-million-square-kilometre Southern Ocean Whale Sanctuary, in which whaling is banned. Despite the vote, however, whales continue to die every year due to whaling activities within the very sanctuary that is meant to protect them. In the 2007 and 2008 season alone, close to 1000 minke whales were 'researched' to death by whalers within the protected area. In addition, the exact number and type of whales being taken through illegal whaling is unknown, although undercover scientists have discovered that aside from species (such as minke) permitted to be killed by Japanese 'scientific' whaling ships, protected species have been taken as well.

FASCINATING FACT

In February 2013, a USA grand jury indicted a now-shuttered Santa Monica area sushi restaurant and two of its chefs for allegedly selling endangered sei whale meat that had been brought in from Japan.

The IWC revised management procedure states that it can reconsider hunting a given whale species if its population recovers to more than 54 per cent of its pre-whaling levels. This opens the door to the resumption of whaling for a number of species that have been protected until recent times. Scientists, however, argue that most whale species were hunted in far larger numbers than those documented by whaling logs and there is no way of knowing accurately how many whales were killed prior to records being maintained. They also argue that while dramatic revelations of massive illegal harvesting and falsified catch numbers have come from Russian biologists, illegal kills by whaling ships from some of the other whaling nations

have never been reported. In essence, therefore, scientists claim that the former numbers of most species would have been far greater than our current population estimates, and that even the so-called 'recovered' species are only a mere shadow of their former selves. Meanwhile Japan, Iceland and Norway claim that whaling is part of their traditional way of life and should be sanctioned. Scientists from these countries also argue that minke whales, which never suffered the heavy losses of other species, are close to full recovery and that sustainable whaling will not harm their populations.

WHALE SCIENTISTS GO UNDERCOVER

In 1993, American scientists Scott Baker and Steven Palumbi travelled to Japan with a secret mission in mind. For a week, the scientists ensconced themselves in a tiny Tokyo hotel where they tested the DNA of whale meat that was being sold in both commercial and consumer fish markets in Tokyo. The whale meat the men were testing should have come from minke whales taken under Japan's research quota, but instead the DNA tests showed that some of the meat had come from protected species such as humpback whales. Since 1993, additional undercover DNA testing has revealed that meat from endangered species such as grey whales is also being sold in Japanese markets. Some of this whale meat is imported to Japan from Norway and Iceland, nations that also continue to defy the international moratorium on commercial whaling.

ESTIMATED PRE- AND POST-WHALING POPULATIONS OF KEY SPECIES

SPECIES	START OF DOCUMENTED WHALING HUNTS	INITIAL NUMBERS	2001 NUMBERS	PERCENTAGE DEPLETED BY
Antarctic minke whale	1921	379,000	318,000	16%
Blue whale	1868	340,000	4730	99%
Common minke whale	1926	258,000	189,000	27%
Eden/Bryde's and Bryde's whale	1909	146,000	132,000	10%
Fin whale	1876	762,000	109,600	86%
Sei whale	1885	246,000	49,090	80%
Sperm whale	1800	957,000	376,000	61%
Southern right whale	1785	86,100	6740	92%
Humpback whale	1664	232,000	42,070	82%

Estimated pre- and post-whaling populations of exploited whales. (Source: Line Bang Christensen, 'Marine Mammal Populations: Reconstructing historical abundances at the global scale', 2006, Fisheries Centre Research Reports 14(9).

GROWTH OF THE ANTI-WHALING MOVEMENT

While it began to dawn on people in the 1970s and 1980s that several whale species were on the verge of extinction, researchers were uncovering new and fascinating information about whales and dolphins. With this increasing knowledge, and growing awareness of the plight of whales, people's perceptions of these marine mammals began to change. International conservation organisations such as Greenpeace, Sea Shepherd, the World Wide Fund for Nature, the Whale and Dolphin Conservation Society, Project Jonah, Whales Alive and others fought – and still fight today – to protect whales from whaling.

Starting with Greenpeace's direct actions against Russian whalers in 1975, anti-whaling organisations have continued to fight to stop the slaughter of cetaceans. For Greenpeace, and Sea Shepherd, this is often achieved by protesters literally putting both themselves and their boats in the paths of the harpoons meant to kill the creatures they are trying to protect. During the 2011 and 2012 whaling seasons, Japanese whaling ships in Antarctic waters made

FASCINATING FACT

On 17 December 2012, the US Supreme Court, citing safety concerns, delivered an injunction to Sea Shepherd USA, ordering them to stay at least 500 yards (almost 460 metres) away from any whaling vessel. Sea Shepherd lawyers subsequently returned to court on 12 February 2013, arguing that the order should be lifted, but their plea was denied, and on 27 February Chief Judge Alex Kozinski of the ninth US Circuit Court of Appeals stated that the activists were threatening the lives of whalers, calling their tactics 'the very embodiment of piracy'.

an early departure from the whaling grounds and headed for home due to the actions of Sea Shepherd activists. In the 2012 season, the whalers took one fin whale and 266 minke whales, less than one-third of their season's quota. The whalers were quoted as saying their season had been 'sabotaged' by the actions of the Sea Shepherd's boats and crew. While there are many who applaud the Sea Shepherd's aggressive tactics, there are also those who feel such actions are unlawful, putting both ships and men at grave risk of death. Paul Watson, Sea Shepherd's founder, remains adamant, however, that he and his crews will continue to target any vessel that is killing whales.

In 2013, the Japanese whaling fleet spent 21 of their 48-day annual hunt trying to avoid Sea Shepherd Australia's vessels, which managed to disrupt them on at least four different occasions. Sea Shepherd vessels also collided with a factory ship and fuel tanker during their attempt to refuel. At the end of the season, the whalers headed home with a meagre catch of 103 minke whales, about 11% of the 935 whales that they had hoped to take. The whalers had also planned to kill 50 humpback and 50 fin whales, but in the end they took none. Sea Shepherd stated that if their American vessels had not been prevented from going to Antarctica by a USA Supreme Court injunction, the catches would have been even fewer, nevertheless they were elated that it was the smallest catch since 1987. The Japanese government, on the other hand, called Sea Shepherd's actions 'unforgivable sabotage', stating that Japanese consumption of whale meat is a cultural tradition and that modern whaling takes place on a sustainable basis. Both sides vow to continue on their current course.

FASCINATING FACT

On 26 June 2013, Australia and Japan squared off at The Hague, in the Netherlands, where the Australian government has taken its case against Japanese whaling in the Southern Ocean to the International Court of Justice, the highest court in the world. The judgements of this so called 'World Court' are binding and there are no appeals. The case has huge ramifications; if Australia wins, Japan will be banned from further whaling in the Southern Ocean. If it loses, Japan may legally continue its 'scientific' whaling programme.

WHALE CONSERVATION IN THE SOUTH PACIFIC

While the effects of whaling in the South Pacific are still being felt, positive changes are taking place in the region to protect the whales and their ocean home. On 17 April 2010, a symposium called 'Ocean Voices … Lessons from the Whales' was held at Auckland Museum and hosted by Tonga's royal patron of whales, HRH Princess Salote Mafile'o Pilolevu Tuita. It brought together members of the South Pacific Whale Research Consortium (SPWRC), dignitaries, scientists and conservationists from New Zealand and other South Pacific islands. International guests at the gathering included Russian biologists Yulia Ivashchenko and Yuri Mikhalev, the latter being one of the scientists who took considerable personal risk in hiding and then making available the true catch data from four Russian whaling ships that were part of the illegal whaling campaign conducted by the former USSR from 1948 to the 1970s (see page 130).

As part of the 'Ocean Voices', members of SPWRC shared the results from some of their research studies carried out in New Caledonia, Tonga, Fiji, Vanuatu, Niue, American Samoa, Samoa, French Polynesia, the Cook Islands and Norfolk Island. While the consortium's primary focus is

In New Zealand, Project Jonah was established in 1974 to campaign against whaling, and subsequently also came up with a mandate to rescue whales that stranded on our shores. Whales that beached themselves were no longer killed to obtain their oil or simply left to die; instead, people mobilised to try to rescue them.

Yuri Mikhalev (former Soviet biologist on whaling ships who revealed evidence of illegal Soviet whaling, right) and Peter Perano (former New Zealand whaler, left) enjoying a day at sea in Kaikoura.

ocean home. One of the highlights of 'Ocean Voices' was the announcement by Tokelau's chief, Aliki Faipule Foua Toloa, of a new marine sanctuary surrounding his islands. Tokelau has now joined 10 other Pacific Island nations that have created whale sanctuaries within their territorial waters. Oceania's marine sanctuaries cover more than 18 million square kilometres, and within them all marine mammals are protected and whaling is prohibited.

The South Pacific Regional Environmental Programme (SPREP), formed in 1982, is another organisation that was established to look after the region's environmental needs. SPREP now includes 21 Pacific Island countries and territories, along with the USA, the United Kingdom, France, Australia and New Zealand. As part of its on-going programme of marine species protection, SPREP developed the 2012–17 Whale and Dolphin Action Plan, with the goal of 'conserving the whales and dolphins and their habitat for the people of the Pacific Island regions'. This comprehensive plan looks at issues such as fishing and pollution, which threaten both the whales and their ocean environment. It also recognises the need for further education, along with the desire to preserve the traditional knowledge and values associated with whales and dolphins. Eco-tourism such as whale-watching is now playing a major role in some Pacific Island communities. However, alongside the benefits these tourism ventures bring comes the need to monitor the industry to avoid undue harassment to whales and other marine creatures, as well as to maximise the ventures' educational value.

The movement to safeguard the home of the whales and dolphins has continued to grow. At the 2012 gathering

on humpback whales, the scientists also collect data on other whale and dolphin species in order to achieve a better understanding of the biology and behaviour of all cetaceans living in the region.

Although the information gained from the SPWRC research has been invaluable, perhaps even more important is the awareness that has been raised throughout the region. Just as the songs of the humpback whale have moved throughout the islands, so too has a renewed song of appreciation spread across the South Pacific community. The human inhabitants of Oceania have listened to the voices from the sea around them and combined lessons from the past with knowledge from the present, joining together to honour the whales and dolphins that share their

Siapo (tapa cloth) from the Pacific islands of Uvea (Wallis) and Futuna, depicting fish and other sea creatures.

of the Pacific Islands Forum, leaders from 15 South Pacific island nations (Australia, the Cook Islands, Federated States of Micronesia, Kiribati, Nauru, New Zealand, Niue, Palau, Papua New Guinea, Republic of the Marshall Islands, Samoa, Solomon Islands, Tonga, Tuvalu and Vanuatu) declared their commitment to the health of the Pacific by formally launching the Pacific Oceanscape, the largest government-endorsed marine management initiative on Earth, covering 38.5 million square kilometres – almost 10 per cent of the planet's surface. During the meeting, the Cook Islands' prime minister, Henry Puna, dedicated his country's newly declared marine park – the largest in the world at more than a million square kilometres – to the Pacific Oceanscape and emphasised the need for Pacific Islanders to 'reclaim their mana [respect] for the ocean'.

Maui's 'Parade of the Whales' has run for over 16 years. Featuring floats, musicians and dancers, the parade opens the annual World Whale Day celebration.

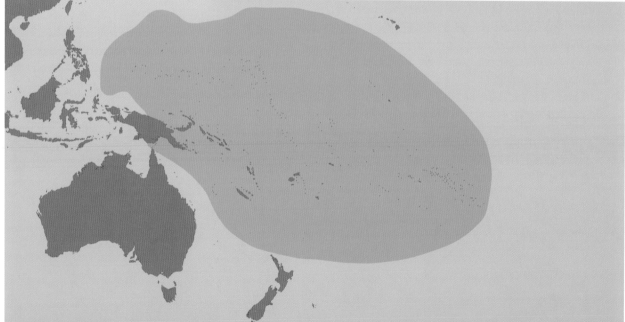

The framework for the Pacific Oceanscape (highlighted in blue) was initiated by the nation of Kiribati. A few hundred years ago, whales entering Kiribati lagoons were surrounded by canoes and driven towards the land, where they would be roped by the tail and dragged ashore by crowds of villagers. Today, the nation works to protect the whales and other marine life within its waters. In 2006, Anote Tong, the president of Kiribati, declared the establishment of the Phoenix Islands Protected Area, which was designated in 2010 as one of the world's largest UNESCO World Heritage Sites.

MARINE MAMMAL PROTECTION IN AOTEAROA NEW ZEALAND

In 1978, New Zealand passed the Marine Mammals Protection Act, which 'provides for the conservation, protection and management of all marine mammals within New Zealand's 4,300,000 square kilometre Exclusive Economic Zone' (EEZ). The EEZ links up with the Southern Ocean Whale Sanctuary to the south of 40°S and, together with the protected EEZs of the South Pacific nations, has in essence created one huge sanctuary in which commercial whaling is prohibited. With the exception of a small area from the tip of the North Island to the Kermadec Ridge, and another off the southern perimeter of Tonga, marine mammals from the islands of the South Pacific to the Antarctic are protected and migrating whale species now have a safe corridor in which to travel between their feeding and breeding grounds.

Six marine mammal sanctuaries have been established in New Zealand since the implementation of the 1978 Act, in which activities that are known to harm marine species are prohibited. One of the sanctuaries is in the Auckland Islands, the main breeding ground of the New Zealand sea lion, while the five others were established to protect Hector's dolphins on the west coast of the North Island, Clifford and Cloudy bays, Banks Peninsula, the Catlins coast and Te Waewae Bay.

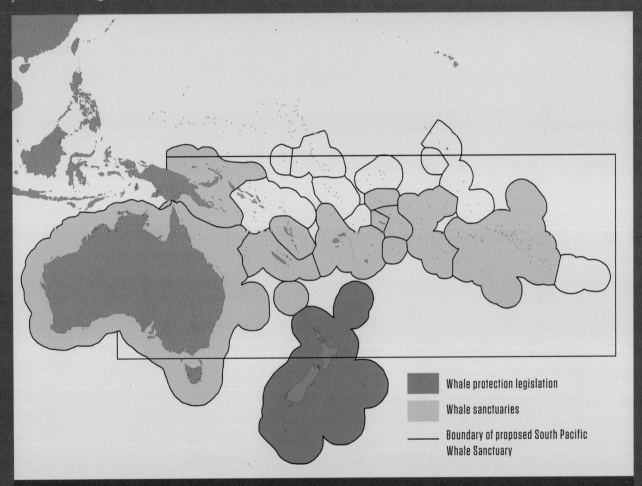

Whale protection legislation

Whale sanctuaries

Boundary of proposed South Pacific Whale Sanctuary

Eleven South Pacific nations and territories (Papua New Guinea, New Caledonia, Vanuatu, Fiji, Tonga, Niue, Samoa, American Samoa, Tokelau, the Cook Islands and French Polynesia) have established whale sanctuaries within their Exclusive Economic Zones. This combined with Australia's sanctuary and New Zealand's whale protection legislation has, in effect, created one large whale sanctuary which covers Australia and New Zealand and stretches across the Pacific to French Polynesia.

THREATS TO CETACEANS

INDIGENOUS AND ILLEGAL WHALING

Aboriginal subsistence whaling is seen by the IWC as separate from commercial whaling and so is not covered by the moratorium. Countries with subsistence whaling agreements are set catch limits, but many opponents argue that this killing is unsustainable, targeting severely depleted species such as bowheads, and in the modern world is totally unnecessary. In the case of illegal whaling kills, protected species such as humpback, fin and sei whales are hunted and their meat is then smuggled into whale-eating countries. Illegal whaling has been undertaken by crews from countries such as Norway, South Korea, Japan, Taiwan and the Philippines. In addition, some countries, such as the Faroe Islands, continue to kill pilot whales as part of their traditional hunt.

BY-CATCH DEATHS

Fishery by-catch of whales and dolphins continues to be one of their biggest threats, despite many new innovations and protective laws. Before the United Nations worldwide ban on large oceanic driftnets was implemented in 1992, it is estimated that more than 300,000 cetaceans were killed each year by these floating death traps, and following the ban the illegal setting of driftnets throughout the oceans has continued to kill many thousands of ocean species. Inshore gillnets and trawl fisheries also continue to be a major threat to small cetaceans such as Hector's and other dolphin species. In addition to nets, whales and dolphins

often become entangled in other fishing gear, such as ropes and buoys attached to crab, lobster and crayfish pots. A number of stranded whales and dolphins are found each year with rope, plastic bags or nets that they have inadvertently swallowed in their throats or stomachs.

BOAT STRIKES

Collision with boats is a particular hazard for species that either make their home in, or migrate through, areas with

While 'scientific' whaling continues to be an on-going threat to some species, numerous whales and dolphins are also dying from other causes – some from permitted aboriginal subsistence whaling, some from illegal whaling, some from 'incidental' catches in fishing gear (like this Hector's dolphin), some from boat strikes, some as a consequence of military sonar or seismic testing and some from pollution.

This whale's tail is a stark illustration of the effect of boat strike on cetaceans, who now share their waters with thousands of commercial and recreational boats.

heavy shipping traffic. A prime example in New Zealand are Bryde's whales living in the Hauraki Gulf, which is utilised by many thousands of boats, including both large commercial vessels and hundreds of small, fast-moving recreational craft. From 1996 to 2012, 43 Bryde's whale deaths were reported in the gulf, with most of these being attributed to boat strikes – an average of two deaths caused by boats every year. Dolphins of many species in New Zealand waters also show evidence of being hit by boats, including damaged dorsal fins and other scarring on their bodies. The issue is a worldwide problem. In the North Atlantic, for example, large numbers of the few remaining critically endangered northern right whales have boat scars on their bodies and many of them die every year from boat strikes.

NOISE POLLUTION

Underwater noise is increasing as a growing number of boats utilise the world's waterways. Just a few hundred years ago, the oceans were full of natural sounds, including the low-frequency pulses of fin and blue whales, humpback whale songs, and the echolocation clicks and high-pitched squeaks and squeals of dolphins. Today, however, boat

WHALE ENTANGLEMENT IN AOTEAROA NEW ZEALAND

Between 2000 and 2012, there were 25 reported whale entanglements around mainland New Zealand, 20 of which involved humpback whales, one a sperm whale, one a southern right whale and three orca. Twenty-two of the whales had become entangled in the floating ropes and buoys from craypots, and while the majority of the entangled animals survived, at least two are known to have died.

In 2003, a Kaikoura man was killed while trying to free an entangled humpback whale off Kaikoura, and since then the Department of Conservation has developed an action plan that does not require human divers to enter the water when they are freeing whales from ropes and nets. Seventeen of the 20 humpback entanglements have occurred along the South Island's east coast (14 off Kaikoura) during the whales' northerly winter migration to their breeding grounds.

traffic and industrial activity have combined to create an
underwater acoustic 'smog' that is something akin to an
visual whiteout, the unnatural noises acting as an auditory
blanket that smothers natural sound. The racket from
hundreds and sometimes thousands of boat propellers is
often combined with depth sounders and other acoustic
gear to produce even more din.

For cetaceans, some of the most debilitating noise comes
from seismic air guns used to map oil and gas deposits and
underwater blasting. Military exercises using high-decibel
mid-frequency sonar – which can cause internal damage to
a whale's ear structures or gas bubbles in its bloodstream,
leading to a form of decompression sickness – are believe
to have caused a number of stranding events.

It is virtually impossible for whales and dolphins to escap
this acoustic invasion into their homes, and species that
once communicated over vast distances may now have th
messages interrupted and may barely hear one another.

GLOBAL WARMING AND MARINE POLLUTION

Global warming poses one of the most serious threats to
whale species and indeed to all of the ocean's inhabitant
One of the potential consequences of the rising ocean

temperatures is a disruption in the production of plankton,
which sits at the base of the marine food web and so forms
the basis for all life in the sea. Once the marine food web
starts to collapse, it will have a domino effect, not only on
whales but on all other species that share this planet.

Pollution is another major enemy. Rubbish is a huge
problem for cetaceans, which often mistake floating junk
for food – a floating plastic bag, for example, is easily
confused with a jellyfish. There is currently so much
rubbish in the seas that literal 'trash islands' have formed
in some areas. Tissue sampling has shown that the bodies
of many ocean species are highly contaminated with PCBs
(polychlorinated biphenyls), mercury and other chemical
pollutants, which may lead to compromised immune
systems. Affected animals are susceptible to viruses
such as the morbillivirus, which was linked to the deaths
of thousands of striped dolphins in the Mediterranean
in 1990–92, and may be responsible for other large and
often unexplained cetacean die-offs that have occurred
elsewhere in the world. Contaminants such as DDT and
other chemicals become more concentrated in animals
higher up the food web, and because many chemicals are
oil-soluble, whales and dolphins store the pollutants in
their blubber and nursing mothers end up releasing them
into their milk, thereby passing them on to their offspring.

Researchers from the University of Montreal and the
St Lawrence National Institute of Ecotoxicology have
shown that beluga whales living in the St Lawrence
River in North America are in danger of being wiped out
through pollution. The population is estimated to have
once numbered around 5000 animals; today, around 500
survive and the dead bodies that wash ashore contain high
levels of heavy metals, PCBs and DDE (a residue from the
breakdown of DDT). Long-finned pilot whales off the Faroe
Islands are similarly contaminated, to the extent that the
islanders – who continue to slaughter around a thousand
of the animals each year as part of their traditional whale

When the Rena *grounded off the Bay of Plenty she held more than 2 million litres of heavy fuel oil and more than 80,000 litres of marine diesel oil, and was carrying 1368 containers, eight of which contained hazardous materials.*

hunts – were advised in 2008 that the whale meat was no longer considered safe for human consumption. If whales are the marine equivalent of canaries in the mine, as some scientists suggest, then many of our oceans are in serious trouble.

Environmental catastrophes such as oil spills pose major threats to all marine wildlife, including cetaceans. Sperm whales are known to have been directly affected by the 2010 BP Deepwater Horizon disaster in the Gulf of Mexico – North America's worst oil spill to date – and a study commissioned in 2011 by the National Oceanic Atmospheric and Administration (NOAA) found that dolphins living in the vicinity are suffering from severe health problems.

New Zealand suffered its own disastrous oil spill on 5 October 2011, when the Greek-owned 236-metre container ship MV *Rena* ran aground on the Astrolabe Reef off the Port of Tauranga in the North Island's Bay of Plenty. The grounding of the large vessel created a huge impact on the people and the environment throughout the region. By 9 October, a 5-kilometre-long oil slick had formed in the waters surrounding the *Rena*, and on 11 October a storm hit the area, causing significant damage to the hull and other parts of the ship that allowed an estimated 350 tonnes of oil to escape. During the same

Dolphins swimming through an oily sea.

Worst affected by the Rena oil spill were seabirds, of which more than 1400 died and more than 400 had to be cleaned of the sticky black goo and subsequently rehabilitated.

FASCINATING FACT

During the *Rena* grounding and subsequent oil spill, more than 300 little blue penguins were rescued and housed in newly built enclosures. The tiny penguins had large appetites, each needing to be fed twice a day with five to seven fish per meal.

storm, at least 70 containers fell off the stricken ship into the sea. By the following day, battered containers and thick oil were washing up on local beaches and the grounded ship had officially become New Zealand's worst maritime environmental disaster.

There is little doubt that whales and dolphins in the area were adversely affected by the oil spill, even though the effects were not as blatant as those of oil-covered birds and seals. Researchers who have examined cetacean species involved in other oil spills around the world have found long-term effects from breathing in the fumes and ingesting toxic oil, including liver damage, ulcerated stomachs and lung damage. During the *Rena* grounding, dolphin species were sighted in waters around the spill, with orca observed coming in close to one of the worst affected beaches, where they traditionally feed on stingrays. Other cetacean species, including endangered blue and humpback whales, are also known to pass through the area. The grounding clearly demonstrated

the enormous threats whales and other marine organisms face today – and will continue to face in the future – from the activities of man.

While environmental disasters such as the 2010 BP Deepwater Horizon spill and the 2011 grounding of the *Rena* are hard to predict, there are many issues we can control. One of the keys to overcoming potential environmental problems is knowledge and education.

EDUCATION

Throughout New Zealand and the islands of the South Pacific, a number of programmes are in place to teach both children and adults about whales and their marine environment. Change is often initiated by education, and one of the best places to start is with children.

One example of just such an education initiative involves Cook Strait Humpback Whale Survey scientists working with LEARNZ, an online educational programme, to enable children in the classroom to engage in a virtual field trip that follows the day-to-day activities of the researchers. During their week-long Internet 'field trip', the students are able to watch the scientists working at sea collecting photographs and skin samples from the whales. They are also able to explore the remains of the former Perano whaling station in the Marlborough Sounds and listen to the whalers describe what it was like in the days

'Save the Whales' by James Sutherland (age 11) of Kaikoura.

of whaling. The educational programme is an opportunity for children to learn not only about a part of New Zealand's history but also about whales, the effects of whaling in the past, and the present impacts that human activities are having on whales and the environment as a whole.

While it is important to educate children – who, after all, will become the whale ambassadors of the future – it is also necessary to inform all people of the issues concerning our oceans. Change is possible, but it takes awareness – along with commitment – in order for it to happen. A number of people in entrenched whaling nations, such as Japan and Norway, are now whale conservationists, and a prime example of changing attitudes can be seen amongst the former New Zealand whalers who now work closely with Department of Conservation scientists to monitor the diminished stocks of migrating humpbacks – the very whales they once hunted and killed. A number of organisations, including the International Fund for Animal Welfare, Project Jonah, Whales Alive, World Wide Fund for Nature, Whale and Dolphin Conservation Society and South Pacific Whale Research Consortium, to name a few, are working tirelessly to ensure that whales have a brighter future, both in the South Pacific and around the world.

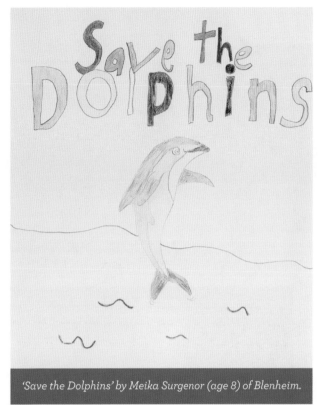

'Save the Dolphins' by Meika Surgenor (age 8) of Blenheim.

Education programmes in the South Pacific are enabling young people to grow up with a better awareness of the ocean ecosystems that surround their island homes.

ISAAC SCOTT:

Young conservationist

My name is Isaac Scott and I am 11 years old. I live in New Zealand and I am passionate about whales. Why? It probably started when I was a year old, and I was in SeaWorld, Florida, and a baby beluga whale followed me up and down the glass. From that moment on, I felt like I had a connection with whales.

Now I am older, I do not advocate places like SeaWorld, where whales and dolphins are kept in glass tanks for human entertainment. A lot of the mammals in aqua parks are taken out of the wild. They are beautiful, graceful and intelligent creatures, and are best seen in their natural environment: the ocean. I have seen whales and dolphins on a few occasions in the wild now too, through whale-watching tours both here in New Zealand and in Australia.

Although we know a lot about whales, we are still learning. I want to study whales when I leave school, and I want to help protect and conserve them for future generations. They are wondrous creatures. My favourite whale has always been the sperm whale, by its looks and the way it hunts out the giant squid.

When I was nine years old I was outraged at the fact that the Japanese were killing whales out in the Southern Ocean under the guise of research. But it wasn't research; it was to kill them for meat! I wanted to join the Sea Shepherd group so that I could stop the killing! I couldn't because I was only nine years old, but I had to do something. I decided to start a petition that would be handed to the New Zealand Government to get the Japanese to stop killing the whales. If I could get enough signatures then someone would have to listen to me. I spent all my school holidays at supermarket doorways, malls and around Wellington Harbour collecting them. I even managed to get all of the Sea Shepherd crew to sign too. In the end, I managed to collect more than 5000 signatures. I also set up a Facebook page, which my parents helped me with. I had some MPs and crew from Sea Shepherd, including Paul Watson and Pete Bethune, to name a few, following my page.

What I was doing caused media attention from newspapers, the radio and even a New Zealand TV station. From the exposure that was given to my cause, I had a phone call from Luit Bieringa, a man who was helping to set up the National Whale Centre that will be based in Picton. He saw my story in the newspaper and asked if I would be interested in becoming the centre's first Youth Patron Member, which I gladly accepted. It even has its own website too, www.aworldwithwhales.com. It hopes to teach people — particularly children — not only about whales but how hunting whales had its place in New Zealand, and how New Zealand found a different way of sustaining itself.

I am an avid follower of the Sea Shepherd group (right), and would say Paul Watson, the founder of Sea Shepherd, is my ultimate hero, totally committed to saving and preserving our oceans and what lives in them. I have met Paul (opposite) on two occasions, once in Wellington and the other in Sydney, Australia, where I was invited to attend a special fundraising event for Sea Shepherd, at which Paul was guest speaker. He spoke to me for about

20 minutes; he is such an inspirational man. He had heard about what I was doing to help the whales and was impressed, and this is when he used a quote I found extremely motivating: 'If the oceans die, then we die.' It's a quote that spurs me on to want to conserve all our marine life.

I really believe we could all make a difference, no matter how small. If we ignore the problem, future generations could be denied the chance to see just how magnificent these creatures really are.

RECONNECTIONS

*Plait the Rope
that Weaves
the Past to the Future*

Pat Hohepa, Discourse at 'Ocean Voices …
Lessons from the Whales' (17 April 2010)

The arrival of whalers in New Zealand waters not only affected the whales, but also altered the lifestyle of Māori and brought about a change in their perspective of, and relationship to, whales. Whales soon ceased to be primarily viewed as kaitiaki (guardians) or a koha (gift) from the sea that provided food and bone; instead, they became commodities with a financial value. In more recent times, a cultural reconnection with whales has been taking place, a reconnection that the indigenous people throughout the South Pacific are carrying with them as they move forward, into the future. In New Zealand, stranded whales are once again viewed as a taonga or koha from the sea. In many cases, such as a 1997 stranding of sperm whales in Golden Bay, New Zealand, stranded whales are viewed as a tohu (sign).

As part of the revival of taonga puoro, the playing of traditional Māori instruments, Hirini Melbourne (musician, composer and linguist), Richard Nunns (musician), Brian Flintoff (master carver), John Haruru (kaumātua), Barney Thomas (manawhenua and Department of Conservation liaison officer) and Robin Slow (renowned artist), along with others of the Golden Bay iwi, were invited to play their instruments in a dawn ceremony at a sacred site on the furthest end of Onetahua (Farewell Spit). One of the reasons for the ceremony was to 'open up' the old trails the whales used to follow. A few days later, five large sperm whales beached themselves at the spot where the instruments had been played. The multiple stranding was acknowledged by the local iwi as Tangaroa's endorsement of the revival of taonga puoro, and as the birth of the legend of Hirini Melbourne calling the whales ashore.

Each of the whales was named, material was removed (teeth and bone) and they were then buried with an appropriate karakia (prayer). Parts of the whales were carved and are in the wharenui Te Ao Marama at Onetahua Marae, which derives its name from Onetahua, the place of shifting sands that cause whales to strand.

In 2008, a special concert, Puhake ki te Rangi (Spouting to the Skies) by Gillian Whitehead, was put on at Te Papa's marae. During the concert, many traditional musical

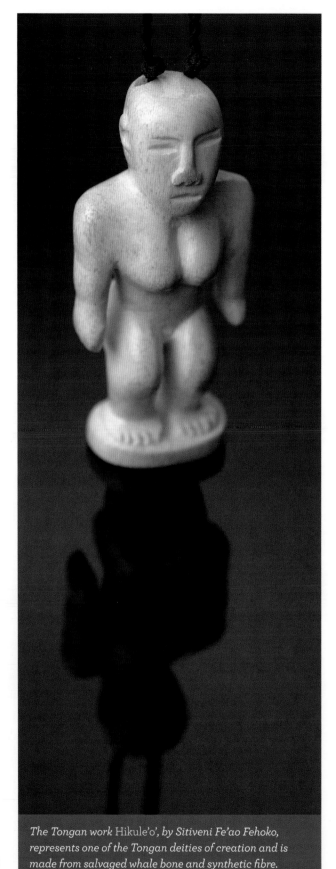

The Tongan work Hikule'o', *by Sitiveni Fe'ao Fehoko, represents one of the Tongan deities of creation and is made from salvaged whale bone and synthetic fibre.*

This work by Robin Slow makes reference to the beaching of five whales at Onetahua. Parapara, the maunga (mountain) of iwi within Mohua (Golden Bay) can be seen, with Taranaki in the distance. On clear days, the mountain can be seen from the area and is important because many of the iwi link back to there. The red ties indicate the death of the whale. At some strandings, when there are large numbers of (particularly pilot) whales, this indicates those that need to be buried and those that need to be returned to the sea. Further references include the blood in the eye, which relates to scientists who removed the eye for their studies, and the buildings on top, which indicate that previously in New Zealand, the economy was based on utilising these whales.

Brian Flintoff describes his nguru: 'It was appropriate for me to carve a flute from the whale named Whaowhia as a mark of the special stranding and the birth of a legend. The carving of the flute also acknowledges [musician] Hirini Melbourne's sharing of his affinity with nature, the wonder and beauty of its sounds, and their relationship with the music, language and mythology that was instilled into him by his elders. The carving is both a whale and a nguru, or flute. The face of the whale is carved in the traditional style used for whales and the flute itself has been named Tutarakauika, which is translated as "the leader of the pod". The nguru's tail flukes are decorated with manaia faces to acknowledge the gift of the enormous strength they hold. The flippers are shown as hands to remind us that these are "whale people" with a senior genealogy to us "human people".' Kura Koiwi: Bone treasures (2011)

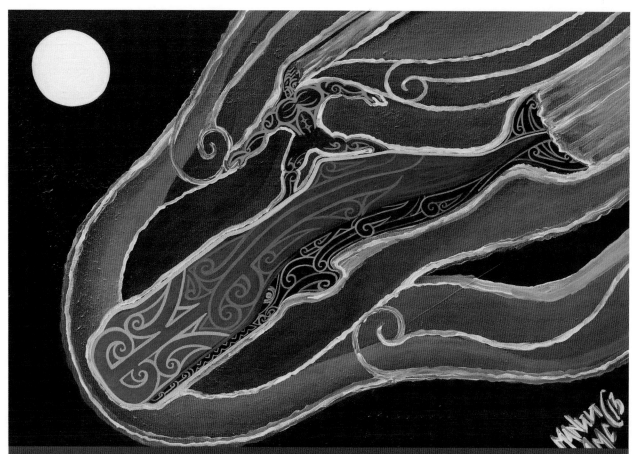

Metua 'Manuu' Tangiatua, Cook Island Māori, is a self-taught artist whose life was changed when he become a guide for Whale Watch Kaikoura. Manuu's experiences at sea with Kaikoura's marine life has been one of the primary inspirations for his work. In this painting, Paikea, the whale rider, Paikea, is encircled within a comet symbolising the Polynesian legend of people coming from the stars to Earth. The three koru around the whale signify baskets of knowledge, while the different moko represent the identities of people and tribes throughout Polynesia.

This pūtōrino by Brian Flintoff was carved from a whale's rib and shaped in the form of a whale.

instruments were played, including the pūtōrino, a special instrument that has two voices: the kōkiri, or male voice, which emerges when the musician blows down into the instrument; and the softer waiata wahine, or female voice, which is created by blowing crossways along the end of the instrument. This renaissance of ngā taonga puoro has revived an instrumental tradition that had been lost for over a century, with many musical treasures lying silent in museums for generations.

Other artists celebrate their reconnection with whales through carvings and paintings. Carvers admire whale bone for its fine wood-like grain and say it offers a quality that no other material can give.

While some people are experiencing a reconnection with whales, others have made a new connection as they have become more aware of their fellow mammals – which live in a foreign environment yet are like us in so many respects. People of all ages express their feelings towards and interest in whales in different ways: some read about or watch documentaries on whales in order to understand them better, while others create poetry, music or art to celebrate their existence. Many thousands of people go on tourism boats to experience whales in their natural environment directly, others join organisations focused on protecting whales. Some march for whales and others

assist cetaceans that have stranded on land. Interactions between humans and whales have taken many forms throughout the centuries, and in some cases those interactions have been initiated by the whales or dolphins themselves.

Ocean Balance, 2007, by New Zealand carver Owen Mapp was made from a sperm whale jaw, pāua shell and goat horn. The two stylised sperm whales represent a healthy and harmonious balance in life and between the land and the sea.

New Zealand artist Rangi Kipa carving whale bone.

ENCOUNTERS
WITH WHALES AND
DOLPHINS

*To the dolphin alone, beyond all other, nature has
granted what the best philosophers seek:
friendship for no advantage.*

Plutarch, *Moralia* (*c.* AD 100)

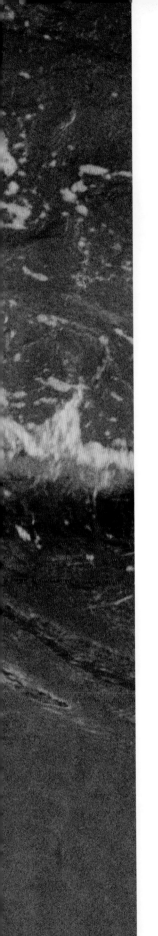

Many centuries ago, Roman lawyer and author Pliny the Younger (AD 61–c. 112) wrote in a letter about a solo dolphin, a bottlenose that befriended a young boy at Hippo, a seaside village in Africa. One day, the boy was swimming when the dolphin appeared and played with the boy, before taking him on his back and carrying him to the shore. Soon the townspeople were wading into the sea to touch and caress the dolphin. The boy and the dolphin continued their close relationship and the boy would 'ride on its back, far out to sea over the sparkling waves'. The story, however, does not have a happy ending. The town's peaceful atmosphere was soon disrupted by hordes of visitors who had heard of the dolphin and wanted to swim with the animal and touch it for themselves. One day the dolphin was found dead, killed by the humans it had befriended and trusted. This story has been repeated again and again since then.

The exact reasons why cetaceans turn to human companionship or to another species remain unknown. Often they are juveniles; in the late 1990s, a young pilot whale that was separated from its pod on the east coast of the USA did not go back to its own kind. Instead, it was adopted by Atlantic white-sided dolphins and was observed living with them for more than six years. Luna, a young orca born in 1999, was separated from his pod in the Pacific Northwest and subsequently turned to humans for companionship. Many different species have become 'human friendly', including beluga, grey and dwarf minke whales, but for some unknown reason the vast majority of solo cetaceans that seek human companionship are bottlenose dolphins. Worldwide, there have been over 90 documented reports of dolphins or whales that have chosen to interact continuously with people, in many cases entirely forsaking their own kind for the company of humans.

AOTEAROA NEW ZEALAND'S SOLO DOLPHINS

Since 1888, New Zealand has recorded no fewer than 12 instances of solo or friendly dolphins. Of these, the most famous were Pelorus Jack, Opo, Maui and Moko, all of which gained worldwide celebrity status.

Pelorus Jack (1888–1912)

From 1888 until 1912, a 4-metre silvery-grey Risso's dolphin escorted passenger ships travelling between Nelson and Wellington. Jack would meet Nelson-bound ships near the entrance to Pelorus Sounds and remain with them for about 8 kilometres to the entrance to French Pass, and then make the reverse trip with the Wellington-bound ship. 'Magic fish follows Nelson steamer' and 'Passengers gasp as bold dolphin shows the way' were just two of many headlines appearing in the daily newspapers.

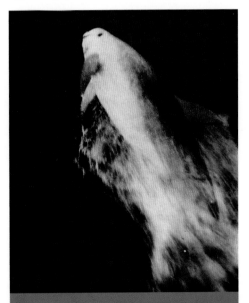

As the Nelson-bound passenger steamer SS Pateena *made its way across Admiralty Bay in 1888, a shout of 'There he is!' would go up, followed by a cheer as a large silvery shape leapt into the air, then another shout, 'It's Jack!'.*

Jack's fame grew and soon he was a star in Aotearoa and around the world. His comings and goings were reported in explicit detail and the Nelson steamer was often crowded with tourists, including famous authors Mark Twain and Frank T. Bullen, who booked round-trip passages just so they could see him. Souvenir postcards with Jack's photo proclaimed 'only fish in the world protected by law'. In 1904 an Order in Council under the Sea Fisheries Act had been put in place to protect the dolphin after someone aboard the SS *Penguin* attempted to shoot him with a rifle.

Ironically, while Jack was being protected, his larger 'fish' cousins living in nearby waters continued to be hunted by whalers. In 1911, the same year Parliament renewed the order of protection for Risso's dolphins, motorised whaling vessels and exploding harpoon guns were introduced in the Marlborough Sounds.

The final sighting of Pelorus Jack was somewhere between November and December 1912. He may have simply died of old age, although some claimed foreign Norwegian whalers had killed him and others that he had been found stranded on a beach after a storm. Risso's dolphins are not common in New Zealand waters and Jack's final disappearance was as much of a mystery as his appearance in the first place.

Opo (1955–1956)

Forty-three years later, another solo dolphin hit the headlines. In the spring of 1955, the quiet seaside village of Opononi had a surprise visit from two adult and one juvenile bottlenose dolphin that crossed the Hokianga River bar and swam into sheltered Omapere Bay. A week later, the two adults suddenly disappeared (possibly shot by locals who mistook them for sharks), and the young dolphin was abruptly bereft of its companions, one of which was probably its mother. For a time, the dolphin merely followed

fishing boats on their rounds, but soon it came closer to the beach and before long the dolphin, now christened Opo, was interacting with people who came to see her.

Soon photographs and stories of Opo started to appear in newspapers all around New Zealand, and before long they had spread to Australia and around the world. Large crowds started flocking to Opononi; the once quiet village had become a tourist attraction. By late February 1956, Opo was a celebrity and the National Film Unit arrived in Opononi to record her performing her repertoire of tricks. By now, Opo was allowing a few children to ride on her back and would often initiate rides with her young friend, 12-year-old Jill Baker.

For the Māori residents of Opononi, the arrival of Opo had a special significance. The full name of the harbour is Hokianga-a-Kupe, the Returning Place of Kupe, and some local Māori believed Opo remained solitary because she was a messenger that had been sent by Kupe. The dolphin was described as 'Opo, the Fish of Peace' during the unveiling of an inscribed stone obelisk memorial to Kupe.

Other locals were not so enchanted with the dolphin's presence, however. Shots were fired at her after she interfered with fishing nets, and soon after she had to be saved from three men who were caught dragging her onto the beach. Radio comments and letters to the Minister of Fisheries expressed concern about Opo's safety, and on 3 March 1956, newspaper headlines read 'Opo Will Soon Have Full Protection'. In a repeat of the Pelorus Jack legislation, Parliament passed a special law to protect any

Two of Opo's favourite playmates were local teacher Mrs Goodson (pictured) and 12-year-old Jill Baker.

dolphin that came to Opononi. But on 8 March, a few days before the law went into effect, Opo disappeared. Her body was later found by a Māori elder wedged among rocks at Te Kauere-a-Kupe, or Kauere Point, and although her death looked suspicious, the exact cause was never known.

News of Opo's death was flashed around the world. She was given a formal burial in front of Opononi's RSA hall, where her grave was littered with floral wreaths, and a statue of Opo was later erected on the town's waterfront. Even today, almost 60 years later, Opo is not forgotten.

Maui (1991–)

In late 1991, a young lone bottlenose dolphin appeared in the small fishing community of Motunau, about 40 kilometres south of Kaikoura. The dolphin was soon following the boats of local fishermen, who nicknamed her Piggy Muldoon. Occasionally, a fisherman would find it very hard to haul his rock lobster pot and, when he finally got it to the surface, he would discover Piggy lying across the top of it. The dolphin remained in Motunau for more than two years and sometimes swam with Aimee Basher, the daughter of one of the fishermen, and also developed a strong rapport with the Bashers' German shepherd. In 1993, the dolphin disappeared from Motunau after a large storm and a short time later she arrived in Kaikoura, where she was rechristened Maui.

Maui remained in Kaikoura for the following two years and became a favourite with locals and tourists alike. Most encounters occurred with people who went out on boats to see her, and, while she had favoured friends, she also interacted with almost any human that entered the water. Wake-riding was one of her favourite pastimes, and if the boat stopped she would swim off and spyhop, staring at the boat until it started moving again and she could resume her fun. She would repeat this manoeuvre until finally the driver gave in and kept going so she could get her wake-riding fix.

In May 1994, Maui travelled north again, this time taking up residence in Queen Charlotte Sound, where she was nicknamed Woody. She continued to follow boats and interact with people in the water, but her encounters became less intense and she spent more time just roaming around. In February 1995, she returned to Kaikoura but had little interest in human companionship and spent most of her time with the local dusky dolphins.

Maui returned to the Marlborough Sounds in late winter 1995, and the following year she was sighted with a calf. The pair would follow boats, but they did not engage in

For some unknown reason, many solo dolphins seem fascinated with boat engines. The fascination shown by solo dolphin Maui bordered on the obsessive, and she would often go up to a boat that had its engine turned off and flick the propeller blade with her nose as if she was trying to start the motor.

Maui enjoyed being rubbed with seaweed, and if people didn't bring some with them, she would swim off, grab a piece and bring it over to them.

Maui often tried to join in with dusky dolphin mating groups, and it was comical as well as a little sad to watch her try to push her large, rotund 3.5-metre shape into a group of lithesome 1.6-metre dusky dolphins.

any prolonged human encounters in the water. Late in 1996, Maui appeared in Wellington, where she escorted one of the local ferries, and as fate would have it, the ferry's captain had associated with Maui in Kaikoura in the mid-1990s. Maui then left Wellington and moved back to Queen Charlotte Sound in late 1997. She remains solitary for the most part but has been sighted with a calf on two occasions, so is obviously having some interactions with other bottlenoses.

Although Maui is one of New Zealand's longest-surviving solo dolphins on record, she has never received the same level of fame as Pelorus Jack and Opo. Moko, New Zealand's most recent solo dolphin, on the other hand, had a huge following of people both nationwide and overseas.

Moko (2007–2010)

Moko was first sighted in March 2007, when he popped up alongside a recreational fishing boat at Mahia in Hawke's Bay. The young bottlenose dolphin soon became a fixture in the bay and local man Joe Hedley gave him the name Moko, after the Mokotahi headland. Moko remained in Mahia for more than two years and became known not only for his friendliness, playing with people and bringing them gifts of fish but also for his mischievous thefts of boogie boards, oars, beach balls and other swim toys. During the summer months, Moko seemed happy with the throngs of people who came to visit him, but some thought

he became bored in winter when nobody was around. In 2008, Moko was involved in the rescue of two pygmy sperm whales, an event reported by media around the world, including a BBC News story on 12 March:

A dolphin has come to the rescue of two whales which stranded on a beach in New Zealand.

Conservation officer Malcolm Smith told the BBC that he and a group of other people had tried in vain for an hour and a half to get the whales to sea.

The pygmy sperm whales had repeatedly beached and both they and the humans were tired and set to give up, he said.

But then the dolphin appeared, communicated with the whales, and led them to safety …

Mr Smith said that just when his team was flagging, the dolphin showed up and made straight for them.

'I don't speak whale and I don't speak dolphin,' Mr Smith told the BBC, 'but there was obviously something that went on because the two whales changed their attitude from being quite distressed to following the dolphin quite willingly and directly along the beach and straight out to sea.'

He added: 'The dolphin did what we had failed to do. It was all over in a matter of minutes.'

In September 2009, Moko left Mahia and travelled north to Gisborne's Waikanae beach, where he remained until late December. By early January 2010, Moko was on the move once again, following the fishing boat *Eskdale* around East Cape, over the Whakatane bar and into the river. Moko quickly settled into Whakatane and soon large crowds were flocking to see him. Moko entertained his human admirers with his cheeky behaviour, as well as with gifts of fish and eels, which he brought to shore and delivered into their hands. He seemed to delight in disrupting boat traffic in the river and would continuously bump a boat's rudder so that steering it became impossible.

In early June, Moko left Whakatane following a fishing boat that was headed north to Tauranga. Later that month, he was sighted in Tauranga Harbour looking 'perfectly healthy and happy', but on 7 July his decomposing body was found washed up on the beach at Matakana Island. The cause of his death could not be determined. Like so many solo dolphins before him, Moko had died well before his time. In his short life, he had touched the hearts and minds of thousands of people, many of whom attended the tangi (funeral) that was held for him on 17 July 2010, when he was buried on Matakana Island.

Like Pelorus Jack and Opo before him, Moko was known around the world. He had his own Facebook page and his exploits were reported in newspapers and broadcast on television world news channels such as CNN and the BBC.

The sighting of a lone whale or dolphin does not necessarily mean that it has forsaken its own kind. A number of whale species live mainly solitary lives, only joining with others on feeding or mating grounds. Many dolphin species live in what is known as fission–fusion societies, groups that are continuously splitting up and forming new alliances. A lone dolphin may therefore simply be scouting for prey or travelling from one pod to another.

Solo dolphins and whales that seek out the company of humans are, however, a different phenomenon. Looking back at both ancient and more recent stories, it is fascinating to see the similarities in their behaviours. Many have initially interacted with boats and have had an obsession with propellers. Most search for objects to 'play' with, either natural objects like seaweed or man-made objects like balls, and the animals' sense of humour is displayed over and over again in the form of numerous mischievous behaviours.

While the presence of a solo dolphin often creates a bond between members of a community who have shared encounters with their new ocean friend, there

One of Moko's favourite pastimes was taking objects such as surfboards and then swimming around with them just out of reach of the owners.

Bill Shortt, one of Moko's early friends, stands beside the memorial dedicated to the solo dolphin that spent 29 of the known 38 months of its life living and interacting with humans at Mahia. The memorial sits on the foreshore in front of the Mokotahi headlands after which the dolphin was named.

are times when the dolphins do not generate goodwill and harmony. People sometimes forget that their new playmate is, in fact, a very large wild creature that does not always behave according to human rules. The dolphin's 'cheeky' behaviour can be extremely disruptive and it is not uncommon for males to become sexually aggressive, which can be very intimidating for the person on the other end of their attentions. In one instance involving Moko in winter 2009, the dolphin would not allow a lone female swimmer who was playing with him to return to shore, to the point where she became exhausted, panicked and had to be rescued.

Thousands of people around the world have experienced encounters with whales and dolphins in the wild, which have left them with indelible and lasting memories. The sight of a dolphin bow-riding, or even a distant whale blow, can generate strong emotions in the viewer. So why do we feel such a strong connection with cetaceans? This is a question that has been asked over and over, but the definitive answer remains a mystery. Is it simply because we are fellow mammals? Or is it because whales and dolphins often approach us in a benign and friendly manner, as opposed to running away or attacking like most other wild animals? In many ways, the question is really two-sided, for the attraction is often reciprocal, with dolphins and whales seemingly relating to us as strongly as we relate to them.

In 1975, underwater expert Wade Doak and his wife, Jan, founded Project Interlock, with the intention of learning more about interactions with wild cetaceans. They invited people to share their experiences with them, and within a few years they had received thousands of letters from people in Aotearoa and around the world. Doak has now written several books filled with the accounts of hundreds of people who have had intense and sometimes inexplicable interactions with whales and dolphins in the wild.

FASCINATING FACT

In March 2011, *Time* magazine named Moko as one of the top 10 animal heroes of all time. Moko won this achievement for the role he played in the rescue of the two pygmy sperm whales that stranded at Mahia in March 2008.

Many solo dolphins display altruistic behaviours such as bringing their human friends 'gifts' in the form of fish and other objects. Here, Moko delivers an eel.

CETACEAN ECO-TOURISM

Since the end of the whaling era, awareness and knowledge about whales and dolphins has grown, and cetaceans have now become symbols of a new appreciation of the environment and of the need to protect its wild inhabitants. This heightened awareness was, and still is, seen in the number of organisations that work to protect cetaceans, as well as in the large number of books, movies and documentaries on whales and dolphins that have appeared during the past four decades.

With the increase in knowledge about whales and dolphins has come a desire to see the animals first hand. This has led to a burgeoning eco-tourism industry that enables people to have their own personal encounters with cetaceans and other marine wildlife. There are many reasons for joining such tours – to learn, to interact directly with the animals or simply to absorb the moment. For many people, the emotional impact of a direct experience leaves a lasting impression and a desire to make a change by becoming actively involved in campaigns to protect whales and the marine environments in which the animals live.

The birth of whale-watching seems to have started with North America's grey whales. As early as 1945, researchers in southern California were carrying out census counts of grey whales migrating along the west coast from feeding grounds in the Arctic to breeding grounds off sheltered lagoons in Mexico. In 1950, the Cabrillo National Monument in San Diego was converted into a land-based whale-watch lookout and attracted almost 10,000 people in its first winter. In the following years, more and more people arrived and the first boat trip to view the whales headed offshore in 1955. Through the 1960s and 1970s, whale-watching from both land and sea spread up the coast of California and into the states of Oregon and Washington. During those years, long-range trips also departed from San Diego to grey whale calving grounds in Mexico.

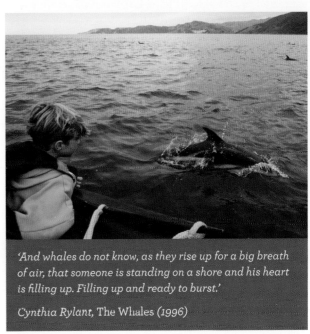

'And whales do not know, as they rise up for a big breath of air, that someone is standing on a shore and his heart is filling up. Filling up and ready to burst.'

Cynthia Rylant, The Whales (1996)

People who encounter wild cetaceans often take home the memory of a life form whose existence has a sense of simplicity and orderliness and, with these, a sense of peace and tranquillity that many humans have lost and perhaps are striving to attain.

In 1971, the Montreal Zoological Society began to offer trips down the St Lawrence River in Canada to see fin, minke and beluga whales, the first whale-watching to occur on the eastern coast of North America. While these early trips had a following and were popular with a number of people, it was really humpbacks that turned whale-watching into a major industry. In 1975, commercial tours to see humpbacks started in earnest on both Hawai'i and in New England, and within a few years whale-watching had become a multi-million-dollar business. In 1981, an estimated 400,000 people around the world went whale-watching, with an estimated direct expenditure of US$4.1 million; by 2008, the figures had risen to nearly 3 million whale-watchers and more than US$870 million of direct expenditure.

Today, whale-watching tours are offered in 87 countries and territories around the world, including countries like Japan, Norway and Iceland that continue to hunt whales. In many places, the tours act as a joint platform for scientists conducting research on the animals being viewed.

CETACEAN ECO-TOURISM IN AOTEAROA NEW ZEALAND

Whale- and dolphin-watching in New Zealand has been popular for many years, although initially sightings were just opportunistic events on scenic boat cruises around the country. In 1985, Ron Bingham launched Black Cat Cruises at Akaroa on the South Island's Banks Peninsula with the aim of showing people both scenic and historic sites in Akaroa Harbour, as well as giving them the chance to view the local Hector's dolphins.

In 1988, commercial whale-watching began in Kaikoura, then a sleepy community with high unemployment whose main connection to cetaceans was rooted in its whaling past. Converging currents off the Kaikoura coast create nutrient-rich upwellings that support diverse forms of marine life. Bachelor male sperm whales use the Kaikoura area as a feeding ground, dining on squid and fish species that live in the deep underwater canyons of the Hikurangi Trench, which plunge to 870 metres a mere kilometre from the coast and to 1600 metres further offshore. In mid-1987, after overcoming many obstacles, ex-fisherman Roger Sutherland and the author finally gained approval for New Zealand's first whale-watching company, Nature Watch Charters, which was launched in January 1988 with the aim of introducing people to the local marine life and at the same time collecting data on the sperm whales and other cetaceans that frequent the area. Soon, hundreds of people began arriving to view the whales and before long people were also swimming with dusky dolphins on many

Nature watch a new venture

After two years of owning the White Morph Restaurant and Gallery, Roger and Barbara Sutherland are venturing out into the world of nature watch expeditions.

Their new business, Nature Catch Charters, will run boat tours from South Bay to Hapuku, taking in the natural playgrounds of local marine and bird life. Mr Sutherland said sperm whales, dolphins, seals and birds such as albatrosses, terns and petrels would be common sights on the trips.

They will also be running land-based expeditions in a four-wheel drive vehicle around the Mt Fyffe area.

Mr Sutherland said the Department of Internal Affairs was helping with the venture, working as booking agents for them in Christchurch.

STILL FINE TUNING

At the moment, Nature Watch Charters is still the process of fine-tuning.

"We won't really be in full swing until about October," he said.

A six-metre rigid hull inflatable is under construction and will be ready for use next month. With a top speed of about 100kmh it will seat 10.

The couple have had the idea for the charter tours for about three or four years but their dreams have only recently become reality after the Ministry of Transport handed the authority of Kaikoura harbour over to the Marlborough Harbour Board. Mr Sutherland said that transfer brought about an easing of boating requirements, allowing up to 6-metre boats to operate within a three kilometre radius of the lighthouse, where previously no boats were allowed.

It has taken the Sutherlands four months to gain approval for the venture and get it off the ground.

FOUND RIGHT BOAT

"Finding the right boat was difficult. We wanted a six metre boat that was going to be safe. I think this one is ideal for the waters around Kaikoura because it's so rigid."

The boat will also contain a hydrophone for listening to marine life underwater.

The tours will venture across the edge of a 700-fathom hole where sperm whales can be seen feeding. Mr

after the southern giant petrel, a seabird which Mr Sutherland used to often see around Kaikoura. He said they might have to sell or lease out the White Morph if and when the charter business takes off.

"I can't see ourselves doing this and Nature Watch Charters as well."

When they bought it, the former Bank of New Zealand building was a bed and breakfast establishment. Mr Sutherland said they spent a year tearing down walls and making structural changes, converting the 85-year-old building into a restaurant and the old bank's strongroom into a wine cellar. The restaurant seats 48 and is gaining quite a reputation for itself.

It has recently been accepted as part of the "Taste New Zealand" restaurant directory, put out by the New Zealand Tourist and Publicity department.

"We've had really good local support for the restaurant," she said.

GALLERY FOR ARTISTS

The gallery, in what used to be the manager's office, is "a conglomerate of quite a few local artists" according to Mrs Sutherland. She said at least 50% was local work.

The White Morph will soon play host to an exhibition of the works of local painter Toos Buurman, whose works have sold well there over the last few weeks.

Before becoming restaurateurs, Barbara and Roger got to know a lot about marine life and the sea. He was a fisherman in Kaikoura for 12 years and she was the photographer for a "whale school" in the San Juan Islands, between Washington state and Vancouver Island in Canada.

There she studied killer whales, then spent 12 months studying sperm whales off the Kaikoura coast. She will act as commentator on the tours.

"We know the area well as far as marine life goes. It's

Roger and Barbara Sutherland bring years of marine experience to their new venture, Nature Watch Charters.

The Kaikoura Star announces the launch of Nature Watch in 1988, New Zealand's first whale-watch company.

of the tours. Within a year, a second boat had to be added to accommodate the large influx of visitors from Aotearoa and overseas.

On 15 July 1989, a Māori-owned company, Kaikoura Tours Ltd, was officially launched by the Minister of Māori Affairs and a second whale-watch operation was introduced in the town. Kaikoura Tours' mandate was not only to give visitors a chance to view whales and other marine life but also to include a Māori cultural perspective. One of the goals of the new company was to provide both current and future employment to Māori youth in an area of the country where employment opportunities were scarce. In the late 1990s, Nature Watch was sold to Kaikoura Tours and the amalgamated companies became known as Whale Watch Kaikoura.

Today, Kaikoura is one of the country's prime destinations for whale-watching, and for swimming with and viewing dolphins and seals. As a result, the town's visitor numbers have risen dramatically, from around 6000 per annum in 1988 to an average of more than 150,000 visitors per annum in the past 10 years.

The success stories in Kaikoura and Banks Peninsula have been echoed around the country, with new whale and dolphin tours springing up in a number of locations, and existing operations increasing their numbers. While successful companies bring economic gains to a community, trips to observe any wild creature offer both benefits and risks to the animals that are being viewed. Some of the benefits come from people learning about and becoming more involved in protecting wild creatures and their habitats, while the risk is the potential disturbance to the animals' normal behaviour patterns.

Toothed cetaceans trying to hunt may be disturbed by the noisy engines of boats, which can interfere with echolocation and their ability to communicate with one another. In addition, all cetaceans have periods of rest and the presence of boats and swimmers during those times is potentially disruptive. High instances of boat strikes are recorded in areas that experience heavy boat traffic from tourist and private vessels.

New Zealand has addressed some of these problems, introducing strict rules regarding marine mammal viewing. No one is allowed to view whales, dolphins or seals commercially without a permit and all operators must follow stringent guidelines, including ensuring that there are no more than three boats within 100 metres of a cetacean and that there is no swimming with whales or with nursery pods of dolphins.

FASCINATING FACT

On 3 February 2013, a giant squid measuring 8 metres in length and weighing around 150 kilograms was found floating in the waters off Kaikoura. This rare discovery is evidence of the existence of this favourite prey of sperm whales in the deep-water canyons off Kaikoura's coast.

In Fiordland, a number of resident bottlenose dolphins exhibit scars from encounters with propellers, reflecting increasing boat traffic in the area.

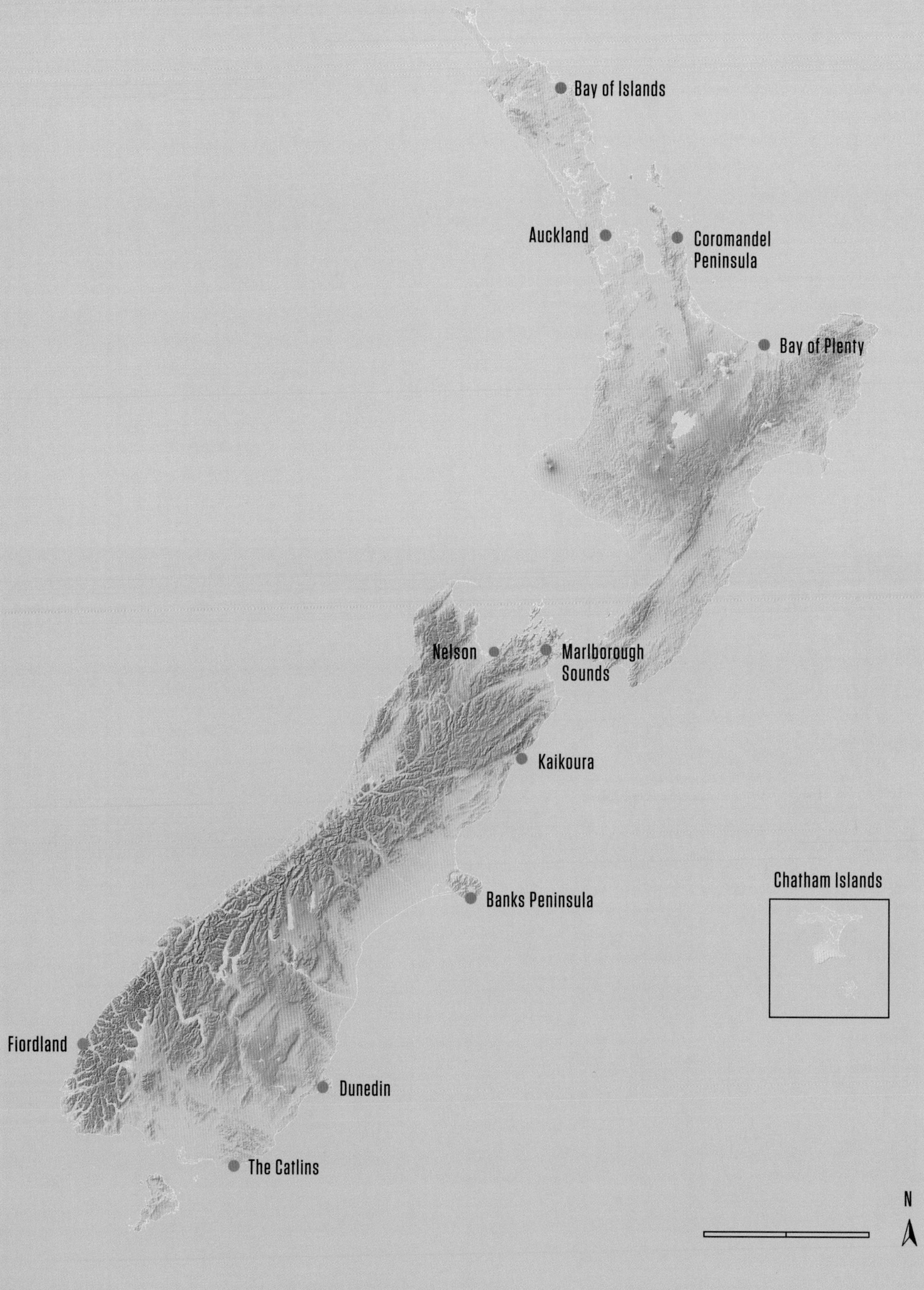

Bay of Islands

Auckland

Coromandel
Peninsula

Bay of Plenty

Nelson

Marlborough
Sounds

Kaikoura

Banks Peninsula

Chatham Islands

Fiordland

Dunedin

The Catlins

N

CETACEAN ECO-TOURISM HOTSPOTS AROUND AOTEAROA NEW ZEALAND

The map opposite indicates the cetacean eco-tourism hotspots around the country and the species you are likely to encounter there.

North Island

Northland/Bay of Islands

Boat tours to observe cetaceans and, in some cases, swim with dolphins can be found at Mangonui, Paihia, Russell, Tutukaka, Whangarei and Ruakaka. Bottlenose and common dolphins are the most prevalent species, although orca, false killer whales and Bryde's whales are also spotted on a fairly regular basis.

Coromandel Peninsula

Year-round trips leave from Whitianga and Whangamata to observe and swim with common dolphins; bottlenose dolphins are also in the area, particularly in late winter and spring, plus the occasional orca.

Auckland/Hauraki Gulf

Boat trips depart Auckland to observe Bryde's whales, bottlenose and common dolphins and, occasionally, orca.

Bay of Plenty

Numerous boat trips are available from Tauranga to observe a variety of cetaceans, including common and bottlenose dolphins, orca, Bryde's and minke whales, and occasional migrating species such as humpbacks. At Whakatane, you can swim with common dolphins and take tours to Whakaari/White Island, where common dolphins and Bryde's whales are often observed, along with occasional sightings of orca, minke and migrating humpback whales.

South Island

Nelson/Golden Bay

While there are not any specific dolphin-watching trips available in the region, dusky, common and, occasionally, bottlenose dolphins, as well as orca, are observed by tour boats carrying people to the Abel Tasman National Park and also by kayakers paddling along the coastline.

Marlborough Sounds

Dusky, Hector's, common and bottlenose dolphins are all seen from locally departing eco-tours, as are orca and occasional migrating baleen whale species.

Kaikoura

Year-round trips leave Kaikoura to watch sperm whales and watch or swim with dusky dolphins. Hector's dolphins are also present year-round and sometimes observed, and there are occasional sightings of common, southern right whale and bottlenose dolphins, as well as pilot whales, orca, humpbacks, southern right whales and blue whales.

Banks Peninsula

Boat trips depart Akaroa daily to watch and/or swim with the Hector's dolphins that live in the harbour.

Dunedin

Wildlife boat tours operating out of Dunedin and the Otago Peninsula may sight bottlenose, Hector's and dusky dolphins.

The Catlins

Boat trips to observe Hector's dolphins, as well as seals, sea lions and penguins, depart from Waikawa Harbour in the Catlins.

Fiordland

Bottlenose and dusky dolphins are often observed by tour boats in both Milford and Doubtful sounds.

PART THREE

WHALES AND DOLPHINS OF AOTEAROA NEW ZEALAND

Schools of porpoises and blackfish are only more animated waves and have acquired the gait and game of the sea itself.

Thoreau

NEW ZEALAND CETACEANS

*When he is about to plunge into the deeps his entire flukes
are tossed erect in the air, and so remain vibrating
a moment till they downward shoot.*

Herman Melville, *Moby-Dick* (1851)

The waters around Aotearoa host no fewer than 42 different species (and three subspecies) of whales and dolphins, or just a little less than half of the world's total of 87 species. Some, like the mighty blue whale, are migratory, passing by annually as they journey between their feeding and mating grounds. Others, like melon-headed whales and hourglass dolphins, are vagrants, normally residing in tropical or polar seas and occasionally venturing into New Zealand's temperate waters. A few species, such as Bryde's whales and various dolphins, reside year-round, and one – the tiny Hector's dolphin, along with its subspecies, Maui's dolphin – is endemic. This diverse range of species includes little-known beaked whales, a few of which have never been sighted alive anywhere in the world and are known only from stranded specimens or their skeletal remains.

HOW TO USE THE SPECIES DESCRIPTIONS

The species described here are organised according to their scientific classification (see page 38), arranged by group, then by family, genus and species. The odontocetes are covered first, including the toothed and beaked whales, dolphins and the one porpoise species that occurs in New Zealand, followed by the mysticetes, or baleen whales.

A fact file is given for each species, detailing maximum size, diet (when known) and presence in New Zealand waters, plus its conservation status according to the International Union for Conservation of Nature (IUCN) Red List. This list categorises species in nine groups according to their risk of extinction, as follows: Not Evaluated, Data Deficient, Least Concern, Near Threatened, Vulnerable, Endangered, Critically Endangered, Extinct in the Wild and Extinct.

In New Zealand, scientists have developed a national status list for marine mammals, designed to complement the IUCN Red List and including the categories Threatened (Nationally Critical or Nationally Endangered), Vagrant, Migrant, Data Deficient and Not Threatened. This information is also provided in the fact file for each resident species under the heading 'New Zealand population'. Sadly, all three of New Zealand's endemic marine mammals are threatened with extinction: the New Zealand sea lion and Hector's dolphin are classed as Nationally Endangered, while Maui's dolphin is Nationally Critical.

Following the fact file is more general information about the species, including the etymology (origin and meaning of its name), descriptions of its appearance and behaviour, its distribution in New Zealand, and threats to the species and conservation related to it.

SPECIES DESCRIBED HERE

ODONTOCETI: TOOTHED WHALES

Worldwide there are no fewer than 73 Odontoceti species. Thirty-four species and one subspecies from four different Odontocete families have been sighted in New Zealand waters. They range from the largest toothed species, the sperm whale, to the smallest, the tiny Hector's dolphin. The species include three Physeteroidea (sperm whales), 18 Delphinidae (oceanic dolphins), 13 Ziphiidae (beaked whales), and one Phocoenidae (porpoise) species.

PHYSETEROIDEA: SPERM WHALES

The three sperm whale species are grouped within the superfamily Physeteroidea. The mighty sperm whale – the largest of all the toothed species – is placed in its own family, Physeteridae, while the two much smaller sperm whale species are placed in the family Kogiidae. The sperm whale is found year-round in New Zealand waters; the pygmy sperm whale is present in offshore waters but seen only occasionally, although it frequently strands on beaches; and the dwarf sperm whale is a rarely sighted vagrant.

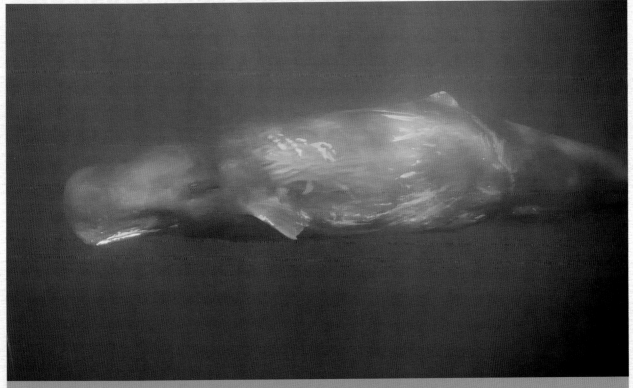

The white skin around the lower jaw of sperm whales may help attract squid and other prey.

Sperm whale, Parāoa

Physeter macrocephalus (Linnaeus, 1758)

Size	Male 14–18 m, 43,000–56,000 kg; female 11–12 m, 13,000–20,000 kg
Gestation	14–16 months
Birth size	3.5–4.5 m, 1000–1500 kg
Calving interval	4–6 years
Lifespan	At least 60–70 years
Teeth	20–26 pairs of large conical teeth, found only in lower jaw
Diet	Medium-sized to large squid species, deep-sea sharks, octopus, rays, and numerous bony fish such as ling, grouper (hāpuku) and orange roughy
IUCN status	Vulnerable
New Zealand population	Resident year-round; Not Threatened

ETYMOLOGY

The sperm whale's genus name, *Physeter*, is Greek for 'blower' and refers to the whale's habit of lying on the surface after a long underwater dive, where it 'blows' (breathes) every 10–30 seconds for at least 10 minutes in order to re-oxygenate its body before beginning another deep dive. The specific name, *macrocephalus*, translates as 'big head' and refers to the whale's extraordinary head, which is up to one-third of the male's body length and a quarter of the female's.

The species' common name came from whalers, who thought the milky, viscous fluid stored in its head looked like sperm. It was soon discovered that this 'sperm' was, in fact, a very high-quality oil that did not smoke or burn when heated to high temperatures. Called 'spermaceti', this oil became very valuable from the late 18th century as a cooking, lighting and heating fuel, leading whalers to target the species in increasing numbers, and in later years as a high-quality lubricant for luxury cars and other types of machinery. For many years, spermaceti was the only oil used to lubricate nuclear submarines and power plants, and it has also been used as a lubricant in spacecraft since it does not freeze in the sub-zero temperatures of outer space.

FASCINATING FACT

Almost ironically, while the spacecraft *Voyager* carried humpback whale songs to outer space, almost all spacecraft were lubricated with oil taken from another cetacean species, the sperm whale.

DESCRIPTION

Sperm whales are the largest toothed whales and display the greatest sexual dimorphism of any cetacean species – the males (opposite, left) are up to a third larger and three times heavier than the females (opposite, right). Unlike most other whales, whose skin is smooth, the skin behind the sperm whale's head is often rippled. The dorsal fin is often little more than a small hump about two-thirds of the way along the back, and the flippers are also relatively small in relation to the body size. The blowhole is also unique in both shape and location, being set far forward and to the left-hand side of the whale's massive box-shaped head.

A sperm whale's lower jaw contains 20–26 pairs of teeth, some of which are up to 23 centimetres long and weigh over a kilogram. It's a mystery why the whales have such formidable dentures when, in fact, they are not really essential in enabling them to catch or eat their prey. Sperm whales feed on squid and fish, which are normally sucked in and swallowed whole; perfectly healthy, well-fed adults have been found with deformed teeth or even no teeth at all in their jaw. The whales may use their teeth to help hold onto large prey, which often fights back – as evidenced by the sucker marks and scars from squid beaks that are found on the heads of many sperm whales.

BEHAVIOUR

Sperm whales spend their time foraging, resting or engaging in social interactions. When foraging, the whales make repeated dives, resting on the surface for an average of 9–13 minutes after each dive. Although most sperm whales feed at depths of 600–1000 metres, they can descend to at least 2800 metres and possibly deeper, in underwater dives that can last more than two hours. The whales normally lift their tail flukes clear of the water prior to making a descent, and often both their descent to depth and ascent back to the surface are nearly vertical. A group of females may spread out over an area of around a kilometre and form ranks when they are foraging, whereas males tend to hunt for prey independently.

Sperm whales eat a lot – the stomach of one individual was found to contain an enormous 12-metre squid weighing more than 200 kilograms, while another's stomach contained no fewer than 25,000 squid remains. The whales spend up to 75 per cent of their time foraging and may dive to depths of 1000 metres and more in search of their favoured prey, tracking them down using

Inside a sperm whale's head is the biggest brain (weighing up to 8 kilograms) of any creature that has ever lived, along with spongy tissue that is saturated with up to 25,000 kilograms of spermaceti oil, used by the whale in echolocation and as a buoyancy aid.

echolocation. Although their main prey is squid, the whales also feed on other fish and shark – they have even been observed preying on deep-sea megamouth sharks, a filter-feeding species that was discovered only in 1976. Some sperm whale populations have learned to feed passively off the long-lines of fishing vessels, sometimes taking the fish off the lines as they are being hauled in and at other times diving down to take prey off the hooks as they lie on the sea floor. Non-food items such as glass balls, rubber boots, charcoal, stones and even oil drums have been found in sperm whale stomachs. The whales have one of the largest gullets of any cetacean species, allowing them to swallow very large prey in one gigantic gulp.

When they are not feeding, the whales' actions vary. Sometimes they will lie quietly at the surface, the males usually alone, although occasionally they are accompanied by females on the breeding grounds. Females and juveniles often cluster close together for several hours. Males will breach on occasion, but most active behaviour is exhibited by the females and juveniles, which may roll over and rub along each other's bodies, or engage in breaching and lob-tailing.

Sperm whales have a highly complex social structure based on age and sex. Females form the core units of these societies, living with their calves in stable 'nursery' pods. Calves nurse for two to three years, although some continue to sneak a drink of milk when they are older. Within the nursery pods, it is not unusual to see one female engaged in communal babysitting duties, watching over a number of calves while their mothers are off feeding. Female calves remain with the core groups as they grow older, but the maturing males leave to form bachelor groups that congregate at feeding grounds like Kaikoura. They are sexually mature by the time they are in their early teens but won't breed before they reach 20–25 years of age and do not return to the mating and calving grounds until they are old enough to be accepted by the females as breeding partners.

NEW ZEALAND DISTRIBUTION

In New Zealand, large bull sperm whales are usually seen in deep offshore waters, while smaller younger males use places such as Kaikoura as year-round feeding grounds. Deep underwater canyons reaching down to 870 metres lie just a kilometre off Kaikoura, and a little further out they drop to 1600 metres and more. These underwater canyons contain deep-water fish and squid species that are favoured by sperm whales, which are therefore attracted to the area. Additionally, the convergence of a cold northward-moving coastal current and a warm southward-moving offshore current creates a constant upwelling, leading to a nutrient-rich sea that supports a rich diversity of marine life. These combined factors create a banquet for all local marine fauna, including the sperm whales.

The majority of sperm whales that visit Kaikoura's rich feeding grounds are young bachelors. These younger males are thought to be between the ages of 14 and 25 years. Females and juveniles have been sighted around Kaikoura on occasion, usually about 15–20 kilometres offshore, but the majority of females around mainland New Zealand are seen further north. Mass strandings of nursery pods have occurred around the North Island on at least five different occasions.

THREATS AND CONSERVATION

Prior to whaling, the worldwide sperm whale population was estimated to be more than a million individuals. The current population is now estimated at around 376,000, or 61 per cent less than the original numbers. Sperm whales have been slow to recover owing to their low reproductive rates (females give birth only every four to six years) and also because the largest males were the whales targeted by the whalers, thus severely reducing the numbers of breeding bulls. Whaling undoubtedly reduced

the overall size of sperm whale males; there were once reports of individual bulls exceeding 20 metres, while today's males seldom reach 18 metres.

Today, hunting for sperm whales is virtually non-existent and substitutes for sperm oil – such as the vegetable jojoba oil – are being used in its place. The species' biggest threats now are from entanglement in fishing gear and ship strikes, along with the effects of marine pollution and global warming.

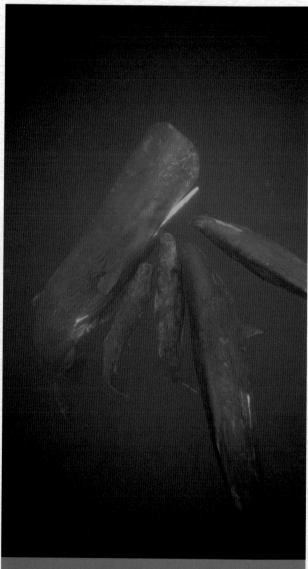

Sperm whale females spend the majority of their lives in tropical or temperate waters at latitudes 0–45°. Older mature bulls visit the females and calves during the breeding season and then depart, migrating towards the poles in order to feed in cooler waters.

Pygmy sperm whale

Kogia breviceps (Blainville, 1838)

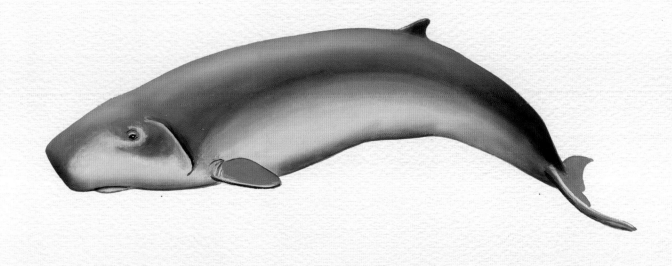

Size	Up to 3.5 m, 350–415 kg
Gestation	9–12 months
Birth size	1.2 m, 50 kg
Calving interval	Unconfirmed
Lifespan	Up to 22 years
Teeth	12–16 pairs in lower jaw
Diet	Mid- and deep-water squid, small crustaceans and fish
IUCN status	Data Deficient
New Zealand population	Resident year-round; Data Deficient

ETYMOLOGY

The origin of the pygmy sperm whale's genus name, *Kogia*, is uncertain but may have come from the 19th-century Turkish naturalist Cogia Effendi, who studied whales in the Mediterranean Sea. Its specific name, *breviceps*, is from the Latin for 'short head'.

DESCRIPTION

The pygmy sperm whale is much smaller than its larger namesake – you could place about five pygmy sperm whales alongside an adult male sperm whale, or two inside its massive head!

Like their relatives, pygmy sperm whales have a spermaceti organ inside a rather odd-shaped head. Their 'short head' appears square from the side but is almost a pointed, conical shape when viewed from above. The blowhole is forward and slightly to the left, and the whale's blow is low and inconspicuous, making it difficult to sight the species at sea.

Stranded specimens have been mistaken for sharks because of their underslung lower jaw and the pigmented crescent-shaped line between their eye and flipper, which resembles a gill slit. The whales have rather unusual and very sharp teeth, which are located in their lower jaw. The 12–16 pairs are thin and curve backwards, and they have no enamel on them.

BEHAVIOUR

Pygmy sperm whales are called *uki kujira*, or 'floating whales', by the Japanese because they lie motionless on the surface of the water. The whales do not lift their tail when diving; instead they merely slip quietly under the surface. They feed on squid, fish and small crustaceans, some of which are bottom-dwellers, indicating that the whales may forage near the ocean floor. The whales are most commonly sighted alone, in pairs or in small groups of less than six individuals.

NEW ZEALAND DISTRIBUTION

While pygmy sperm whales are rarely seen at sea, more than 300 have stranded along New Zealand's coast, particularly around Mahia Peninsula. The large number of single stranded females and cow–calf pairs indicates there is a breeding population of pygmy sperm whales living in deep waters off the North Island's east coast, although why so many strand is unknown. In March 2008, Moko, the well-known solo bottlenose dolphin that resided off Mahia Beach at the time, rescued two pygmy sperm whales from stranding by leading them out to open water through a narrow channel (see page 176).

THREATS AND CONSERVATION

While these whales were never targeted by commercial whaling, they have been hunted in small coastal fisheries in the Caribbean, Japan and Indonesia. Pollution is also clearly a problem, as a number of stranded whales have been found with blockages caused by plastic bags. Pygmy sperm whales are difficult to spot because of their propensity to lie motionless on the water's surface, which makes them vulnerable to boat strike.

This pygmy sperm whale stranded at Raglan in February 2013.

Dwarf sperm whale

Kogia sima (Owen, 1866)

Size	Up to 2.7 m, 135–275 kg
Gestation	12 months (estimate)
Birth size	1 m, 40–50 kg
Calving interval	Unconfirmed
Lifespan	Unknown
Teeth	13 pairs in lower jaw, up to 3 pairs in upper jaw
Diet	Mid- and deep-water squid, fish and crustaceans
IUCN status	Data Deficient
New Zealand population	Vagrant

ETYMOLOGY

Pygmy and dwarf sperm whales were considered to be a single species until 1966, when they were separated into two species. The dwarf sperm's specific name, *sima*, is from the Latin for 'flat-nosed'.

DESCRIPTION

The species has a smaller body and an even shorter head than the pygmy sperm. Dwarf and pygmy sperm do, however, share many of the same characteristics: their body shape and colour patterns are almost identical; they both have blunt-shaped heads containing a spermaceti organ; and they both possess a low, underslung jaw and a gill-shaped mark behind the eye that extends to the front of the flipper. These similarities still cause a great deal of confusion when trying to separate the two species. The dwarf sperm whale's dorsal fin is slightly taller and further forward on its body, but this is hard to ascertain at sea.

BEHAVIOUR

Like sperm whales, dwarf sperm whales use echolocation to locate their prey, which includes squid, crustaceans and fish. When startled, both dwarf and pygmy sperms employ a squid-like tactic: they have a sac in their lower intestine filled with about 12 litres of a dark reddish-brown liquid, which they expel into the water, creating a thick cloud that may deter predators or conceal their escape. Dwarf sperm whales have been spotted in aggregations of up to 10 animals.

NEW ZEALAND DISTRIBUTION

Although pygmy sperm whales have been known in New Zealand waters since the early 1870s, there were no records of dwarf sperms until 1981, when an individual (initially thought to be a pygmy sperm) stranded off Waiwera Beach in northern New Zealand. When scientist Alan Baker reviewed the photos and measurements of the whale, he realised it was more likely to be a dwarf sperm whale, and so a major necropsy was completed on the animal.

Results from the necropsy and comparisons of the whale's skull with those of other dwarf sperms from around the world proved conclusively that the whale was in fact a dwarf sperm whale – the first ever recorded in New Zealand. The species is clearly a rare visitor to our shores, preferring to live in warmer temperate and tropical oceans.

THREATS AND CONSERVATION

Like pygmy sperm whales, dwarf sperms have a propensity to swallow plastic bags and other ocean debris, which may result in intestinal blockages and, eventually, death. Some dwarf sperm whales have been observed as by-catch in driftnet fisheries, while others have been harpooned in commercial fisheries in the Caribbean and Indian oceans. The whale's habit of lying on the surface has also led to occasional boat strikes.

When startled, dwarf and pygmy sperm whales expel reddish-brown liquid, possibly to deter predators.

DELPHINIDAE: OCEAN DOLPHINS

Worldwide, there are no fewer than 36 species in the family Delphinidae, known as oceanic dolphins. These include a few dolphins that also inhabit freshwater habitats, as well as six species, such as orca and pilot whales, which although called whales are actually dolphins.

The members of Delphinidae differ greatly in terms of their size and colour patterns, and while some have a pronounced beak, others have hardly any beak or none at all.

Eighteen Delphinidae species (and one subspecies) are known in New Zealand waters. One species, Hector's dolphin (and the subspecies Maui's), is endemic, eight others occur year-round and are considered to be residents, and the remaining nine appear only rarely and are considered to be vagrants.

All members of the Delphinidae species have conical teeth, although the size and number of the teeth varies from species to species.

Hector's dolphin, Tutumairekurai, Tūpoupou

Cephalorhynchus hectori (van Beneden, 1881)

Size	1.2–1.6 m, 45–60 kg
Gestation	10–11 months
Birth size	60–75 cm, 9 kg
Calving interval	2–4 years
Lifespan	Up to 25 years
Teeth	24–31 pairs
Diet	Small fish such as cod and mullet, squid and crustaceans
IUCN status	Hector's dolphin *Cephalorhynchus hectori hectori* Endangered; Maui's dolphin subspecies *Cephalorhynchus hectori maui* Critically Endangered
New Zealand population	Endemic year-round resident; South Island *Cephalorhynchus hectori hectori* c. ≥ 7000, Nationally Endangered; North Island *Cephalorhynchus hectori maui* < 100, Nationally Critical

ETYMOLOGY

The genus name *Cephalorhynchus* roughly translates as 'beak head', while the specific epithet *hectori* and common name were given in 1881 by Belgian palaeontologist Pierre-Joseph van Beneden to honour naturalist Sir James Hector, who first described the dolphin in 1870 but did not, at the time, recognise it as a new species.

Although van Beneden recognised that Hector's dolphin was a new species when he described it in 1881, he initially placed it in the wrong genus, calling it *Electra hectori*. In 1885, Sir James Hector corrected the error and gave the dolphin its correct genus name of *Cephalorhynchus*.

To Māori, they are known as tutumairekurai, meaning 'special ocean dweller', or tūpoupou, which means 'to rise up vertically'.

DESCRIPTION

Like the other three members of *Cephalorhynchus*, Hector's dolphin has a short, rounded body and a head that lacks a noticeable beak. It has a distinctive rounded dorsal fin, which has given rise to its nickname of Mickey Mouse dolphin thanks to the fin's resemblance to the cartoon character's ears. The dolphins are a light grey overall, with black fins and face mask, and a cream belly. They are the smallest of the Delphinidae, measuring just 1.2–1.6 metres long.

BEHAVIOUR

Despite the fact that Hector's dolphins are endemic to Aotearoa, little was known about them for almost 90 years after their recognition as a species. In 1973, Alan Baker and W.F.J. Mörzer-Bruyns published a small paper on Hector's, but the first extensive research on the species did not take place until 1977, when Baker began a field study in Cloudy Bay on the northeast coast of the South Island. The results of his pioneering research were published the following year, making available new information on Hector's biology, behaviour, distribution and diet. Since 1977, a number of New Zealand and overseas researchers have amassed a considerable amount of information on these tiny dolphins.

Female Hector's dolphins are slightly larger than the males (around 10 centimetres longer) and normally have their first calf between seven and nine years of age. The calves are usually born in spring or early summer and stay with their mother for one to two years. A calf will nurse for up to a year, although they start catching fish at around six months of age. Females have a new calf every two to four years, so may produce only six calves in their 20–25 year lifespan.

Hector's are normally sighted in small groups of between two and 10 individuals, although larger congregations are also observed – especially during mating periods, when one or more groups will come together. The dolphins become quite active during periods of sexual activity, swimming rapidly and often leaping on and over each other. They are commonly sighted close to shore in murky waters, particularly near river mouths, and do not appear to be great travellers. Their average range is around 30–60 kilometres, although many dolphins do not even travel that far; movements in excess of 100 kilometres are rare.

Hector's are not deep divers and normally feed in waters of a depth of 80 metres or less. When feeding, the dolphins dive for an average of 90 seconds and dine on arrow squid and small fish species such as yellow-eyed mullet, kahawai, sprats and red cod.

MAUI'S DOLPHINS

In the late 1990s, a study of a tiny population of Hector's dolphins off the west coast of the North Island was undertaken and yielded some interesting results. Researchers discovered that these dolphins have longer and wider skulls and are about 10 centimetres longer than their southern counterparts, and subsequent DNA tests revealed that they are genetically distinct from the South Island population, from which they may have been separated for at least 15,000 years. The DNA results, combined with the morphological differences between the North and South island specimens, were enough for the North Island Hector's population to be classified as a separate subspecies, named *Cephalorhynchus hectori maui*, or Maui's dolphin. Research so far indicates that Maui's have a similar biology to South Island Hector's in their breeding, life span and social organisation. Their prey is slightly different, with Maui's feeding on grey mullet along with kahawai, arrow squid and small crustaceans.

NEW ZEALAND DISTRIBUTION

More than 7270 Hector's dolphins live around the South Island's coastlines. An estimated 5400-plus live along the west coast, around 1800 are found along the east coast and fewer than 90 occur along the south coast. While most Hector's are observed in small groups, larger social gatherings of 30–60 animals are also seen on occasion.

The current range of the Maui's dolphin subspecies extends from north of Kaipara Harbour almost to New Plymouth, but the majority are sighted along a 40-kilometre stretch of coastline between Manukau Harbour and Raglan. The tiny, critically endangered population is estimated at 55 adult animals (one year and older), of which half may be female with perhaps 14–20 of them being breeding females. Maui's tend to live closer to shore than their South Island cousins, making them even more susceptible to being trapped in inshore fishing nets. This led to the establishment of the West Coast North Island Marine Mammal Sanctuary in 2008.

From 2010–12, the DNA of 44 individuals was sampled within Maui's distribution range, with four of the dolphins being genetically identified as South Island Hector's dolphins (two living females, one dead female and one dead male). It is unknown if the two living female interlopers remained in Maui's territory, but if dolphins from the South Island begin to extend their range and breed with the North Island population, it will ultimately affect Maui's subspecies status. On the up side, a new influx of dolphins into Maui's critically endangered population could help increase numbers

Scientist Alan Baker gave the North Island subspecies the name Cephalorhynchus hectori maui *in honour of Maui, the legendary Māori hero who discovered, or 'fished up', the North Island.*

and give the dolphins in the North Island a better chance at survival. Additionally, two South Island Hector's (one dead female neonate and one living male) were identified from the southwest coast of the North Island, outside the presumed range of both subspecies. The combined findings demonstrate long-distance dispersal of over 400 kilometres by South Island Hector's as well as the possibility of an unsampled population of dolphins living off the southwest coastline of the North Island.

THREATS AND CONSERVATION

A number of researchers have been studying Hector's dolphins since 1984, during which time they have examined every aspect of their lives, from their biology to population numbers. Elisabeth Slooten and Steve Dawson's work at Banks Peninsula during the 1980s confirmed the fear, first raised by Alan Baker, that Hector's were particularly susceptible to being caught in inshore gillnets, including amateur set nets, as they tend to occur close to the coast in waters less than 100 metres deep. The researchers discovered that at least 230 Hector's had reportedly drowned in gillnets in the Pegasus Bay/Canterbury Bight area between 1984 and 1988, out of an estimated population at the time of around 740. As a result of this research and following extensive public consultation, the then Minister of Conservation, Helen Clark, established New Zealand's first marine sanctuary, covering over 1170 square kilometres, around Banks Peninsula on 20 December 1988.

In October 2008, the Banks Peninsula sanctuary was increased to 4130 square kilometres, extending along 390 kilometres of coastline between the mouths of the Rakaia and Waipara rivers and 12 nautical miles (22 kilometres) out to sea. Within it, amateur set netting is limited year-round and restrictions are placed on commercial gillnetting and trawling.

On 23 October 2008, four new marine mammal sanctuaries were established to aid in the protection of Hector's and

The dark colour of this Hector's dolphin calf indicates that it is newly born; its colour will lighten within a few months. The new calf is about 60–75 centimetres long, almost half the size of its mother, and by the time it is a year old it will be almost as large as an adult.

Early shore-based whalers rather unglamorously called Hector's dolphins 'puffing pigs' in reference to their somewhat dumpy body and short, puff-like blows.

Maui's dolphins, one in the North Island and three in the South Island. Within the North Island sanctuary, established to protect the Maui's dolphin population and extending along the west coast from Maunganui Bluff in Northland to Oakura Beach in Taranaki, gillnets are prohibited within 7 nautical miles (13 kilometres) and trawling within 2 nautical miles (3.7 kilometres) of the coast. In 2012 the North Island sanctuary was extended 90 kilometres from Oakura Beach south to Hawera.

In the South Island sanctuaries, gillnetting is banned within 4 nautical miles (7.4 kilometres) of the east and south coasts, and 2 nautical miles (3.7 kilometres) of the west coast, during three months of summer, while trawling is prohibited within 2 nautical miles of the east coast.

Because Hector's are such slow breeders, they remain at risk even with the establishment of marine mammal sanctuaries to protect them, and it will be centuries before they are able to come even close to their original estimated population of 20,000–30,000, if in fact they can do so at all. For this reason, the continued ban of gillnetting in the sanctuaries is essential to the survival of Hector's as a species, and especially to the Maui's subspecies, which is classified by the IUCN as Critically Endangered. Like all cetacean species, the dolphins are also at risk from pollution and global warming, which ultimately could pose even greater risks to their survival.

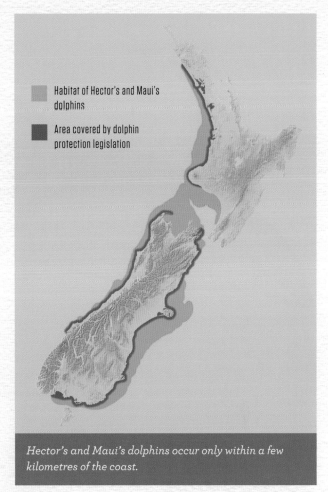

Habitat of Hector's and Maui's dolphins

Area covered by dolphin protection legislation

Hector's and Maui's dolphins occur only within a few kilometres of the coast.

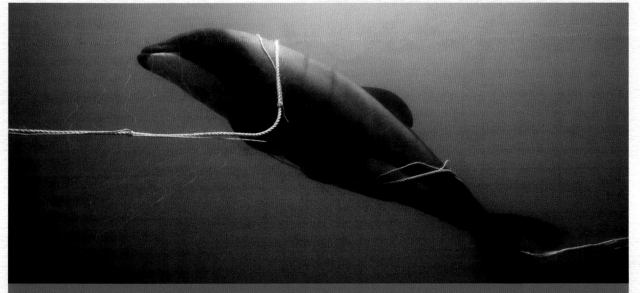

Gillnets are an efficient way of catching fish – the fish swim into the nets and are caught by their gills, hence the name. They are common as set nets in New Zealand, and hang in the water with floats along the top and weights on the bottom to keep them vertical. The size of the net's holes, or mesh, varies depending on the fish being targeted; larger holes catch bigger fish and allow smaller ones to swim through. Unfortunately for dolphins, almost any gillnet they swim into becomes a fatal trap, especially as they can't swim backwards.

Dusky dolphin

Lagenorhynchus obscurus (Gray, 1828)

Size	1.6–1.8 m, 70–85 kg
Gestation	11 months
Birth size	90–102 cm
Calving interval	2–3 years
Lifespan	26–30 years
Teeth	60–65 pairs
Diet	Small schooling fish, mid-water prey such as squid, lanternfish and hake
IUCN status	Data Deficient
New Zealand population	Resident year-round; commonly sighted; Not Threatened

ETYMOLOGY

John Gray is credited with being the first person to name the dusky dolphin based on his description of a specimen captured in 1827 near the Cape of Good Hope, South Africa. In 1828, he assigned the name *Delphinus obscurus* to the dolphin but, in actual fact, the same species had already been described and named *Delphinus superciliosus* in 1826 by René Lesson and Prosper Garnot, who had collected a specimen off the coast of Tasmania. During the 73-year period between 1826 and 1889, the dolphin was actually assigned nine different genus names. Finally, in 1889, the species was classed in its current genus *Lagenorhynchus*, whose name is Greek and means 'bottle beak'. The dolphin's name changes may not be finished, however, as recent DNA analysis may lead to a further reclassification of the species. The specific name, *obscurus*, Latin for 'dark' or 'indistinct', refers to the dusky colour pattern on the cetacean's body.

DESCRIPTION

Dusky dolphins generally have a dark grey back and white throat and belly, with two white stripes running along the sides of the body from the dorsal fin to the tail. They have a short, almost indistinguishable beak and measure about 1.6–1.8 metres in length.

BEHAVIOUR

The dusky dolphin is a southern hemisphere species found mainly along the coastlines of South Africa, South America and New Zealand, with occasional sightings and strandings in South Australia and Tasmania. Duskies are renowned for their acrobatic displays: they spyhop and tail-slap, they leap high and low, and they do somersaults and back flips, all at high speed. When one dolphin starts, others usually follow suit, and their actions are often repeated over and over again. It all appears to be for pure enjoyment, but while some actions may be performed for fun, others are known to be part of the dolphins' feeding and mating strategies.

Dusky dolphins live in fission–fusion societies, which simply means that group structures are almost constantly in a state of flux. Individuals will easily break from one group to join or fuse with another, and group sizes may vary depending on the behavioural need. Small foraging groups, for example, may be best for finding prey, while larger groups may be more efficient for herding and capturing prey. Mating groups may only have five to 15 dolphins in them, while larger groups of dolphins may gather together when they are resting. The most stable relationship is that between a mother and her calf, and they will often join with other mother–calf pairs to form nursery pods.

The diet of dusky dolphins varies depending on where they live. In some areas of New Zealand, such as Admiralty Bay in the Marlborough Sounds, they primarily feed on schooling fish such as pilchards, sprats and yellow-eyed mullet. In other areas, such as Kaikoura, the dolphins occasionally 'snack' during the day, but their main feeding occurs at night, when squid and fish species such as lanternfish migrate vertically from deep water closer to the surface of the sea.

The dusky dolphin's specific name, obscurus, *Latin for 'dark' or 'indistinct', refers to the dusky colour pattern on the cetacean's body.*

Dusky dolphins live in fluid group structures, with group sizes changing depending on the behavioural need.

New calves have foetal folds and floppy dorsal fins that stiffen up within a few days of birth.

Dusky dolphins reach sexual maturity between seven and eight years of age, and appear to be promiscuous – individuals have been observed mating with three to four different partners in less than five minutes, and they have also been seen engaging in sexual activities with common dolphins. Social sex occurs year-round, but mating for the purpose of reproduction happens only at certain times, with most calves being born between late October and mid-January.

In New Zealand, females give birth to calves measuring 90–102 centimetres after a gestation period of around 11 months. Like all cetaceans, dusky mothers are protective and will herd their young away from any perceived danger. Calves remain close to their mother's side for the first few months and may nurse for up to 18 months, although they learn to forage and hunt for themselves at a much younger age.

NEW ZEALAND DISTRIBUTION

In New Zealand, dusky hotspots include the east coast of the North Island from East Cape to Cape Palliser, and from Marlborough Sounds down to Oamaru on the east coast of the South Island. While dusky dolphins are primarily a coastal species, they are often seen offshore, particularly around the Chatham Islands. The southernmost New Zealand record comes from the discovery of three stranded dolphins on Campbell Island, reported by Alan Baker in 1977.

Duskies are gregarious and are often seen with other species such as Hector's (pictured in foreground), common, bottlenose and southern right whale dolphins, as well as pilot whales.

THREATS AND CONSERVATION

Although dusky dolphins are taken for human consumption in Peru, incidental capture in trawling and purse seine fisheries is the major threat worldwide to both duskies and other dolphin species. In New Zealand, the number of fisheries-related deaths is comparatively low, but careful monitoring of the fishing industry, aquaculture and tourism needs to be on-going to ensure their effects on the species are kept to a minimum. Again, as with all dolphin species, pollution and climate change continue to be major threats.

Hourglass dolphin

Lagenorhynchus cruciger (Quoy and Gaimard, 1824)

Size	1.6–1.9 m, 80–94 kg
Gestation	Unknown
Birth size	1 m
Calving interval	Unknown
Lifespan	Unknown
Teeth	52–70 pairs
Diet	Fish, squid and crustaceans
IUCN status	Least Concern
New Zealand population	Vagrant

ETYMOLOGY

Originally described from a fairly crude drawing made at sea in the early 1820s, the species' specific name, *cruciger*, is Latin for 'cross-bearing', a reference to the criss-cross pattern on its body, which, in truth, only vaguely resembles an hourglass.

DESCRIPTION

Early whalers and sealers called the dolphins sea skunks because of their black and white stripes. Males are slightly bigger than females, with noticeably larger and more hooked dorsal fins.

BEHAVIOUR

The dolphins are generally seen in small groups of four to 10 individuals, although larger groups of 40 or more have also been observed. Almost nothing is known about their biology or feeding habits, aside from the fact that they consume squid and small fish species.

Hourglass dolphins are most often sighted in subantarctic and Antarctic waters at latitudes 45°S to 66°S. They are truly oceanic and have never been regularly studied owing to their remote distribution, although they are enthusiastic bow riders when a boat appears in their vicinity. The dolphins are known to associate with long-finned pilot whales as well as fin whales – whalers often used their presence as an indication that fin whales were nearby.

NEW ZEALAND DISTRIBUTION

There have been sightings southeast of the Chatham Islands and around the subantarctic islands. The species is also known on the mainland from three separate specimens that stranded along the east coast of the South Island, the most recent in 2010 at Flea Bay, Canterbury.

THREATS AND CONSERVATION

To date, no real conservation problems have been identified.

Short-beaked common dolphin

Delphinus delphis (Linnaeus, 1758)

Size	Male 1.7–2.4 m; female 1.6–2.2 m; both up to 200 kg
Gestation	10–11 months
Birth size	80–90 cm
Calving interval	1–3 years
Lifespan	At least 25 years
Teeth	82–110 pairs
Diet	Small schooling fish and squid
IUCN status	Least Concern globally; Mediterranean sub-population Endangered
New Zealand population	Resident year-round; commonly sighted; Not Threatened

ETYMOLOGY

The short-beaked common dolphin is the species most familiar to early people of the Mediterranean, a fact reflected in its scientific name, *Delphinus delphis*, which is simply a combination of the Latin and Greek words for 'dolphin'. Common dolphins are found throughout the world and exhibit subtle differences in their colour and shading patterns. In 1995, a new species, the long-beaked common dolphin *Delphinus capensis*, was recognised and a subspecies of *D. capensis*, known as the tropical common dolphin, may in time also come to be recognised as a separate species.

DESCRIPTION

Short-beaked common dolphins have a black back and white belly, with an hourglass-shaped marking on the flanks that is creamy tan towards the front and grey towards the tail. The beak is long with small, sharp teeth.

BEHAVIOUR

Like their long-beaked cousins, short-beaked common dolphins are sometimes observed as a sea of dolphins, appearing in large, active groups that may number in the thousands. As in many dolphin societies, short-beaked commons live in fission–fusion societies, frequently moving from one group to another. Mating occurs regularly, and while new calves are sighted year-round most births take place in late spring and early summer.

Short-beaked commons are gregarious and have been observed in New Zealand associating with at least seven other cetacean species, including Bryde's, minke and sei whales, as well as striped, bottlenose, dusky and Hector's dolphins. Commons often forage in mixed-species groups that include seabirds such as terns, shearwaters and Australasian gannets, other cetacean species and sharks. The prey of short-beaked commons differs depending on the areas in which they are feeding but usually consists of arrow squid, jack mackerel and anchovies. Their foraging methods include individual high-speed chases and fish whacking, where a dolphin hits a fish with its tail, flipping the prey into the air and then capturing it when it hits the water. Coordinated feeding strategies include cooperative herding, where dolphins drive a fish school towards another group of dolphins, effectively trapping the fish between them. The dolphins have also been observed herding fish into a tight ball by swimming round and round them in ever-decreasing circles, trapping them against the surface.

NEW ZEALAND DISTRIBUTION

Short-beaked common dolphins occur in both coastal and oceanic New Zealand waters, with a population residing year-round in the Hauraki Gulf. The dolphins are seen all around the North Island, with the most common sightings along the northeast coastline. In general, they come closer to shore in the summer, the exception being in the Bay of Islands, where the dolphins tend to move offshore during summer.

Large pods of short-beaked commons have been sighted in Cook Strait, while both large and small groups are observed in the Marlborough Sounds and along the west coast of the South Island. In the summer months, a group of commons consistently visits Kaikoura on the South Island's east coast, where they interact and even mate with dusky dolphins (see page 202), occasionally resulting in hybrid offspring.

The beautifully marked short-beaked common dolphin is found in temperate oceans throughout the world.

THREATS AND CONSERVATION

Between 1998 and 2008, 115 short-beaked common dolphins were reported as incidental by-catch in New Zealand fisheries. Over 85 per cent of the deaths occurred within the jack mackerel trawl fishery, mostly on vessels carrying Ministry of Fisheries observers. The exact number of unreported casualties is unknown, but extrapolation based on the number of vessels without observers indicate there may have been as many as 600 deaths. Stranding records during the same 10-year period revealed that at least another 24 short-beaked commons died as a result of net entanglement. Continuing research on numbers is underway, and on the impacts of tourism, pollution, fisheries by-catch and other threats that may affect the New Zealand population.

In 1965, Napier's Marineland caught Daphne, the first common dolphin to be displayed in their aquarium. Over a period of 43 years the centre took at least 74 dolphins into captivity. The two longest-surviving common dolphins, Kelly and Shona, arrived at Marineland in 1974; Shona died in 2006 at the age of 36, while Kelly died in 2008 at the age of 37. Napier's Marineland closed the year Kelly died and has not reopened since.

Common bottlenose dolphin

Tursiops truncatus (Montagu, 1821)

Size	2.4–4 m, 270–550 kg
Gestation	11–12 months
Birth size	1–1.3 m, 15–30 kg
Calving interval	At least 3 years
Lifespan	40–50 years, possibly longer
Teeth	36–54 pairs of stout, pointed teeth
Diet	Mostly fish species but some squid and crustaceans
IUCN status	Least Concern globally; Mediterranean sub-population Vulnerable; Fiordland sub-population Critically Endangered
New Zealand population	Resident year-round; commonly sighted; Fiordland and Bay of Islands sub-populations Nationally Endangered

ETYMOLOGY

The common bottlenose dolphin's genus name, *Tursiops*, means 'dolphin-like', from the Latin *tursio* (dolphin) and the Greek suffix -*ops* (meaning 'appearance'), while the specific name, *truncatus*, means 'shortened' and may refer to the dolphin's flattened teeth. The species' common name is thought to have originated in England and reflects the similarity between the dolphin's snout and the rounded form of an old-fashioned gin bottle. Scientists now recognise two separate bottlenose species: the common bottlenose, found worldwide in tropical and temperate seas, including New Zealand waters; and the Indo-Pacific bottlenose (*T. aduncus*). The Burrunan dolphin, a southern form of *T. truncatus* found off Victoria, Australia, has also been proposed for species status.

DESCRIPTION

The bottlenose has been well known for centuries and in recent times has often had a starring role on television, in movies and at marine aquariums. For most people, it is the archetypal dolphin, with a short, rounded snout and hooked dorsal fin. Common bottlenoses are large dolphins, reaching lengths of 4 metres, and have a light to dark grey back and a pale cream belly.

BEHAVIOUR

As with many other cosmopolitan cetacean species, the common bottlenose chooses its prey and feeding techniques depending on the local habitat, with some hunting alone and others working together in cooperative groups. Some common bottlenoses have been observed fish whacking – hitting fish with their tails to stun or kill them – while others have been seen strand-feeding, where they work together to push schools of fish onto mudflats and then partially strand themselves to feed on their helpless prey. Bottlenose dolphins have also been known to fish cooperatively with humans. In the Brazilian village of Laguna, the dolphins drive fish towards men standing with nets in shallow water, and then move away so the men can throw their nets. The dolphins then dine on any fish that escape.

Female bottlenoses are sexually mature at between 5 and 13 years, while males reach sexual maturity slightly later, at 9 to 14 years. Most females give birth in spring and summer after a gestation of around 12 months, and they nurse their calves for at least a year. Even though a calf is nursing, it will eat solid food and many begin to forage for prey at four to six months. Female bottlenoses have given birth in their late 40s and they are known to live into their mid-50s. Males appear to have a slightly shorter lifespan of up to 50 years.

Common bottlenose dolphins are very gregarious and often associate with other cetacean species, particularly pilot whales. They have also been known to interbreed with other dolphin species, both in captivity as well as in the wild. The dolphins enjoy the company of humans and will seek out boats or swimmers to 'play' with, and they are the species most often credited with coming to the aid of humans in trouble at sea. Just one example occurred in 2004, when four lifeguards swimming off New Zealand's northeast coast were approached by a large shark believed to be a great white. Common bottlenoses surrounded the swimmers and stayed with them for over 40 minutes while they swam safely together to the shore. The species has also been known to help other cetaceans.

It has long been known that common bottlenoses are extremely intelligent, and their ability to be trained easily has been utilised – some would say exploited – by humans, who have captured them for display in marine parks, as well as for research and military purposes. Since the 1960s in the USA, the common bottlenose's acute echolocation sense has been made use of to search for mines and enemy divers. Military sonar is virtually useless on the sea floor, which is littered with different objects, whereas bottlenose dolphins can distinguish mines from other clutter buried in the mud.

NEW ZEALAND DISTRIBUTION

In New Zealand waters there are both inshore and offshore sub-populations of common bottlenoses. The pelagic, or offshore, dolphins range further from land and have been sighted as far north as the Kermadecs and as far east as the Chathams. These bottlenoses are often sighted in the company of false killer or pilot whales. There are three

This young bottlenose calf will retain a strong bond with its mother for several months and may continue to nurse for up to three years or until its mother becomes pregnant again.

regional subpopulations inhabiting coastal waters around New Zealand: one in the North Island, one in the Marlborough Sounds and the third in Fiordland. Each of these groups is isolated from the others both physically and genetically, and they rarely, if ever, interact with one another.

The North Island's population resides year-round along Northland's east coast and in the Hauraki Gulf, with infrequent sightings reported as far away as Manukau Harbour on the west coast. The dolphins are mainly seen along a 500-kilometre stretch of coast between Tauranga and Doubtless Bay, and many of the dolphins utilise the waters in the Bay of Islands. Scientists have conducted studies on bottlenose in the Bay of Islands since 1993 and have identified 408 individuals, some that are frequent users, some that are occasional visitors and some that are merely transients. The research has shown that the abundance of dolphins using the area has declined in recent years at a rate of 5.8–7.5 per cent per year. This decline has coincided with an increase in adult deaths and a higher than normal mortality rate for calves, with an estimated 42 per cent of calves dying before reaching one year of age. The exact reasons for the downturn in the number of dolphins using the Bay of Islands is unknown but may be a combination of a change in habitat use, perhaps due to a lack of prey or increased boat traffic, along with the high adult/calf mortality which is resulting in a low rate of recruitment.

Bottlenose dolphins sighted within the South Island's Marlborough Sounds are part of a larger coastal regional population that ranges to the west coast as far south as Westport and to the east coast as far south as Cloudy Bay. The total number of dolphins within this population is estimated at over 380, with 335 individuals being photographically identified between 1992 and 2005.

The bottlenose dolphin has a short, rounded snout.

As at the Bay of Islands, it appears that some dolphins in the sounds are frequent visitors with high site fidelity, while others only occur occasionally. Research indicates that an average of 195–232 individuals enter and leave the Marlborough Sounds throughout the year and, while the average group size is around 12, larger pods ranging from 80 to 170 individuals have also been observed. The Marlborough Sounds obviously play an important role in this regional population's home range, with some dolphins migrating in and out of the region every seven days or so.

In the third population, found in Fiordland, the dolphins are further divided into three discrete local units or groups. One group ranges along the northern Fiordland coast from Martins Bay to Charles Sound and is estimated to include around 47 individuals; the second group consists of around 56 individuals and occurs in the area formed by Thompson and Doubtful sounds; and the third group, estimated at 102 individuals, lives in the waters formed by Breaksea and Dusky sounds.

In 2011 the IUCN changed the status of the Fiordland regional population of the common bottlenose dolphin to Critically Endangered.

THREATS AND CONSERVATION

While common bottlenose dolphins are abundant and not considered to be endangered worldwide, their national status in New Zealand was changed in 2009 to Nationally Endangered based on the relatively small national population and high calf mortality. In 2011 the IUCN changed the status of the Fiordland regional population to Critically Endangered.

In Fiordland, possible causes of the decline include disturbance and boat strikes, increased freshwater discharge into Doubtful Sound from the Manapouri hydroelectric power station, and reduced availability of prey as a result of overfishing in Fiordland.

The Doubtful/Thompson sounds unit declined by as much as 34–39 per cent over the 12-year period leading up to 2006, with the greatest increase in mortalities, 62.5 per cent, being among both calves and juveniles (aged one to three years). Members of this group also exhibit four times the amount of skin lesions found in other populations, and have smaller calves and a more restricted calving season. If the current trend continues, the entire Doubtful/Thompson group could disappear within the next 50 years.

The northern Fiordland group, which includes Milford Sound in its range, experiences particularly intense pressure from tourist boats. The dolphins now appear to avoid or vacate the sound during the hours of the day and seasons of the year when activity is at its peak. This small group is also declining due to high levels of calf mortality, and like the Doubtful/Thompson sounds dolphins, individuals display scars caused by boats and at least one calf is known to have been killed by a boat strike.

The isolated nature of Breaksea and Dusky sounds has attracted fewer tourist boats, resulting in less pressure from boat traffic on this dolphin group, although tourism is now expanding into the area. Prey depletion within these sounds may be a concern, however, as a number of commercial vessels fish within the area.

To protect the Fiordland sub-population, new marine reserves were established in 2005. In Milford and Doubtful sounds, a voluntary code of practice was established for tour boat operators in 2006 to reduce their impact on the dolphins, and in 2008, 'dolphin protection zones' were established to limit boat activity. These zones extend 200 metres out from the shore in areas that are thought to be most frequently used by the local dolphins.

Killer whale

Orcinus orca (Linnaeus, 1758)

Size	Male to 9.8 m, 10,000 kg, dorsal fin to 1.8 m; Female to 8.5 m, 7500 kg, dorsal fin to 90 cm
Gestation	15–18 months
Birth size	2.1–2.6 m, 160–180 kg
Calving interval	4–6 years
Teeth	Large, slightly curved; 20–28 pairs
Lifespan	Male at least 50–60 years; female at least 80–90 years
Diet	Extremely varied: numerous fish species (including sharks and rays), cephalopods, penguins and other seabirds, marine turtles, seals, and other cetacean species
IUCN status	Data Deficient
New Zealand population	Resident year-round; < 250 individuals; Nationally Critical

ETYMOLOGY

The killer whale was first described in 1558 by Swiss naturalist Conrad Gesner from a stranded specimen, and was given the binomial name *Orcinus orca* by Carolus Linnaeus in 1758. The genus name *Orcinus* means 'the kingdom of the dead', while the specific name *orca* may be from the Latin word *orc*, used to describe a large fish, whale or sea monster.

Killer whales received their common name from whalers, who called them 'whale killers' when they observed them preying on other cetacean species. The name ultimately turned around and they became known as killer whales. For many years, the killer whale's reputation was as fearsome as its name, but that has since changed as researchers have learned more about them and the public has become acquainted with them in marine parks and through documentaries, movies and whale-watching. As a result, many people now prefer to use the more benign-sounding common name of orca for the species, although both names are used interchangeably.

DESCRIPTION

Killer whales are a truly cosmopolitan species, ranging throughout every ocean from the tropics to the poles. The males (bottom) are noticeably larger than the females (top) and possess a distinctive dorsal fin, which may reach 1.8 metres high. Orca are the biggest Delphinidae species; one of the largest known male killer whales was 9.8 metres long and weighed more than 10 tonnes, while the largest recorded female was 8.5 metres and weighed more than 7 tonnes. Orca have a striking black and white colour pattern with a distinctive white eyepatch and a greyish white saddle patch behind the dorsal fin, which is unique to each individual.

BEHAVIOUR

Long-term research in the Northwest Pacific found that, in those orca populations, individuals normally remain within their same matrilineal family pods throughout their entire lives and this is quite possibly true of orca around the world. It appears that female killer whales give birth to their first calf at about 12–15 years of age after a gestation period of 15–18 months. They breed into their 40s and will have an average of five calves during their fertile years. Calving intervals may vary but seem to average five years. Although weaning begins at around a year, mothers may nurse for two years or more. Other pod members – both male and female – assist in rearing the young calf. Females live longer than the males and are one of the few cetaceans known to experience menopause.

Killer whales are supreme predators that adapt their choice of prey and hunting methods to suit their living conditions. They are powerful, fast swimmers, reaching speeds in excess of 55 kilometres per hour, and often hunt in packs. The whales' hunting techniques are so efficient that it is almost impossible for their chosen prey to escape. Author and killer whale expert Erich Hoyt remarked in 1984 that 'The only way to survive an encounter with an orca is reincarnation.' New Zealand orca feed mainly on fish. Eagle, short-tailed and long-tailed rays appear to be favoured, although orca do feed on other rays, as well as blue, thresher and even hammerhead sharks. New Zealand killer whales have also been observed feeding on some dolphin species, including false killer whales.

In the Southern Ocean, four forms of killer whale, known as types A, B, C

An orca with a freshly caught stingray in its mouth.

In March 2010, a pod of around eight killer whales attacked a group of false killer whales in front of a boatload of startled tourists in the Bay of Islands. At least three of the smaller Pseudorca were rammed from below, and a mother and calf pair were killed and devoured by the orca pod.

and D, have been recognised, based on their appearance. Although each type is different in terms of size and colour, the major distinguishing feature is the eye patch, which varies in shape and size. While the range of the four types sometimes overlaps, they do not associate and, in fact, appear to avoid each other. It is expected that further research, including DNA testing, will lead to some of the forms being assigned full species or at least subspecies status. Types B, C and D killer whales have been seen in New Zealand waters, although sightings are extremely rare and these forms are considered to be vagrants here.

Orca in New Zealand waters physically resemble the type A orca, which looks like a 'typical' killer whale, with a striking black and white body and medium-sized white eye patch. In Antarctic waters, type A orca have been observed feeding on minke whales, although they are known to take other prey, including different fish species. Type A orca are predominately observed in open waters and appear to move north during winter months.

Type B orca are smaller than type A and have a larger eye patch, which is slightly yellow in most individuals. The majority of the dark parts on the body are a medium grey colour, with a darker slate-grey patch – the 'dorsal cape' – stretching from the forehead to just behind the dorsal fin. Type B orca are normally found near the shore and feed primarily on seals.

Type C orca are the smallest and are usually found in larger groups. Their colour resembles that of the type B animals, but the eye patch has a distinctive forward slant. Type C orca often occur in dense ice and feed primarily on fish such as Antarctic toothfish and cod although they also prey on penguins and seals. Both types B and C whales live close to the ice pack for much of the year, and diatoms in the water are thought to be responsible for the yellowish tinge found on the white parts of their bodies.

Type D orca have been recognised only recently based on photographs of a 1955 mass stranding that occurred at Paraparaumu, New Zealand, and six additional sightings at sea that have occurred since 2004. Type D orca have the strong black and white colour pattern of type A whales, but their dorsal fins are shorter and narrower and curve backwards. The extremely small eye patch is situated behind the eye on a head that is quite bulbous, similar to that of pilot whales. Little is known about their diet, although they have been observed around long-line vessels targeting Patagonian toothfish.

Type A

Type B

Type C

Type D

Based on a correlation of sightings, type D orca appear to range between 40°S and 60°S; they were sighted at Campbell Island, 54°S, in December 2009.

The foraging strategies of killer whales vary depending on their prey and its habitat. Type B whales, which feed mainly on seals in the Antarctic, have been observed lunging onto ice floes to capture their prey. They have also been seen 'hiding' beneath the floes and waiting for the seals to enter the water, as well as swimming beneath the floes to rock them violently, thereby causing their intended prey to slide into the sea. They even work together in groups to create waves to wash seals off floes.

Elsewhere in the southern hemisphere, in Patagonia and around the Crozet Islands, killer whales have learned to surf onto the beach with a wave, catch a seal pup and then slide back into the sea as the wave recedes. Also in Patagonia, the whales will wait for high tide and then swim through narrow channels into shallow inshore areas where seal pups are playing.

Fish-eating killer whales also work cooperatively to chase and catch their food. When foraging on schooling fish, some will circle the fish while others feed, and then they will switch roles. Some populations have been observed releasing bursts of bubbles in order to herd the fish into a tight ball, then whacking the fish ball with their tails to kill or stun their prey.

NEW ZEALAND DISTRIBUTION

There appear to be three sub-populations of killer whales living in New Zealand: one pod that inhabits North Island waters, one pod in South Island waters and a third that travels between the two islands. Researcher Ingrid Visser and her team have identified at least 117 individual killer whales in Aotearoa, distinguishing them by their distinctive saddle patches, as well as by scars and nicks or unusually shaped dorsal fins. The estimated population of fewer than 250 resident orca in New Zealand has led to a national classification status of Nationally Critical.

THREATS AND CONSERVATION

The most significant worldwide threats facing killer whales are pollution and habitat disturbance caused by underwater noise and conflicts with boats. In some populations, such as in the USA's Northwest Pacific, a decrease of prey species has been linked to a decline in numbers. In New Zealand waters, some killer whales display scars from boat strikes.

The IUCN currently classifies the status of killer whales as Data Deficient. This is because of uncertainty over whether they should actually be split into several species rather than the current single species. If the species is subdivided, it is expected that the IUCN ranking for some of the new species will be changed.

Orca dives are relatively short, averaging anywhere from four to ten minutes.

Long-finned pilot whale

Globicephala melas (Traill, 1809)

Size	Male to 6.3 m, 2300 kg; female to 5.7 m, 1300 kg
Gestation	12–15 months
Birth size	1.7–1.8 m, 75 kg
Calving interval	3–5 years
Lifespan	Male 35–45 years; female to 60 years
Teeth	16–26 pairs
Diet	Cephalopods (mainly arrow squid and octopus)
IUCN status	Least Concern globally; Mediterranean sub-population Data Deficient
New Zealand population	Resident year-round; commonly sighted; Not Threatened

ETYMOLOGY

The long-finned pilot whale's genus name, *Globicephala*, comes from the Latin *globus*, meaning 'globe' or 'ball', and the Greek *kephale*, meaning 'head'; while its species name, *melas*, is from the Greek *melanus*, meaning 'black'. The common name, pilot whale, refers to the belief that one member of the pod is a pilot that leads the group. Pilot whales are also referred to as potheads because of their bulbous heads, and – along with false killer, pygmy killer, melon-headed and killer whales – as blackfish. While each of the blackfish species carries the name 'whale', they are actually all members of the dolphin family, Delphinidae.

DESCRIPTION

Compared to short-finned pilot whales (see page 217), the tapered flippers of long-finned pilots are longer (a quarter to a fifth of their body length), they are slightly larger overall and their skull is narrower, although these distinguishing characteristics are almost impossible to discern when the species are sighted at sea. The biggest clue to their identity lies in their range. Long-finned pilot whales are normally found in deeper, cool to cold waters offshore, foraging for squid (their favourite prey) in depths of 600 metres and possibly more.

BEHAVIOUR

Long-finned pilot whales live in social groups consisting of small mixed pods that gather into large aggregates of up to several hundred animals. The smaller pods are thought to be matrilineal, forming around adult females and their offspring. They are relatively stable and consist of adults, juveniles and mother–calf pairs. DNA studies have shown that several calves within one pod are fathered by the same male, which appears to come from outside that group. It is believed these outside males enter a pod for a period of time and mate with a number of different females during their stay. Females continue to give birth into their late 30s and will sometimes nurse their last offspring for up to 10 years or longer.

NEW ZEALAND DISTRIBUTION

Long-finned pilots are frequently sighted in New Zealand waters. They are also the species that most frequently mass strands – in New Zealand alone, more than 10,000 pilot whales have stranded in the past 100 years. The world's largest whale stranding occurred in 1918 in the Chatham Islands, when around 1000 long-finned pilot whales beached, while in 1985, 450 stranded on Great Barrier Island (Aotea Island).

THREATS AND CONSERVATION

In the northern hemisphere, long-finned pilot whales are killed as by-catch in long-line, trawl and gillnet fisheries, and in the Faroe Islands the indigenous annual hunt (called the *grindadráp*) continues despite worldwide condemnation, killing an average of 850 pilot whales per year. This number may decline, however, as heavy metals such as cadmium and mercury have been found in the local pilot whales, as well as pollutants such as DDT and PCBs, and the human consumption of their meat is consequently being discouraged. There are few reports of southern hemisphere pilot whales being taken as incidental by-catch, and here – as in the northern hemisphere – the species is classed as Least Concern.

Long-finned pilot whales feed mainly at night and rest, socialise or travel during daylight hours.

Male long-finned pilot whales can be distinguished from the females by the size and shape of their dorsal fins, which are much larger and broader than the curved fin of the female.

Short-finned pilot whale

Globicephala macrorhynchus (Gray, 1846)

Size	Male 6.2 m; female 5.1 m, both up to 3600 kg
Gestation	Around 15 months
Birth size	1.4–1.9 m, 55 kg
Calving interval	3–6 years
Lifespan	Male 45–50 years; female over 60 years
Teeth	14–18 pairs
Diet	Mainly squid, plus some fish and octopus species
IUCN status	Data Deficient
New Zealand population	Vagrant; occasionally sighted

ETYMOLOGY

The species name, *macrorhynchus*, means 'enlarged beak or snout' and may refer to the appearance of the whale's melon, which looks a bit like a large snout when viewed from the front.

DESCRIPTION

Short-finned pilot whales are almost identical in appearance to long-finned pilot whales (see page 215), but, as their name indicates, their flippers or fins are slightly shorter (about one-sixth of their body length). They also have fewer teeth, a shorter body and a shorter, broader skull.

BEHAVIOUR

Short-finned pilot whales are usually seen in large schools or pods of 50 or more individuals. When travelling they may swim abreast of each other, in a line several kilometres across. There are close matrilineal associations within the pod, and mature females outnumber mature males. It is thought that males leave the natal pod as they mature but that females may remain within their birth pod for life. Female short-finned pilot whales give birth up to 40 years of age and appear to have an extended nursing period with their last calf that may last more than 10 years.

Short-finned pilots are deep divers, known to reach depths in excess of 800 metres in search of squid, their favoured prey. Research in the Canary Islands has found that individual pods communicate using their own discrete repertoire of tonal calls – something that is likely to be true of short-finned pods around the world. In New Zealand, scientists attached acoustic tags to 12 short-finned pilot whales in 2010 in order to log sound, depth and orientation. They discovered the whales used tonal calls for communication, even when foraging at depths as great as 800 metres, although the call output and duration did decrease as the whales dove deeper.

Females give birth into their late 30s and will sometimes nurse their last offspring for up to 10 years or longer.

NEW ZEALAND DISTRIBUTION

Short-finned pilot whales were first recognised in New Zealand waters in 1977. Although their range overlaps slightly with long-finned pilot whales, the short-finned prefer warm temperate and tropical waters, and are usually seen around the top half of the North Island and further north.

THREATS AND CONSERVATION

While New Zealand short-finned pilot whales are relatively unthreatened, other populations are at risk from gillnet fisheries, pollution, hunting and capture for marine aquariums. Like their long-finned cousins, the whales are susceptible to mass stranding.

Short-finned pilots are deep divers, known to reach depths in excess of 800 metres in search of squid, their favoured prey.

False killer whale

Pseudorca crassidens (Owen, 1846)

Size	Male 6 m, female 5 m, both up to 2000 kg
Gestation	14–16 months
Birth size	1.6–2 m
Calving interval	Up to 7 years
Lifespan	Up to 60 years, possibly longer
Teeth	14–24 pairs of thick, curved teeth
Diet	Fish and squid
IUCN status	Data Deficient
New Zealand population	Resident year-round; occasionally sighted; Not Threatened

ETYMOLOGY

The false killer whale was first described from sub-fossil remains found in Lincolnshire, England, in the mid-1840s, when it was named *Phocaena crassidens*. As there had never been any live sightings of the species, the whale was thought to be extinct until a number stranded in Germany in 1861. The genus name *Pseudorca*, or 'false orca', was later assigned to the whale based on its similarity to the killer whale (see page 211) in terms of skull shape and in the size and shape of the teeth. The species epithet, *crassidens*, meaning 'thick-toothed', was retained.

DESCRIPTION

False killer whales do not look very much like their orca cousins. Their slender body is almost entirely black, with a greyish-white blaze on the throat and chest, their dorsal fin is small and slightly hooked, and their tapered flippers have a distinctive hump or bulge on their leading edge.

BEHAVIOUR

False killer whales have an extensive range in warm temperate and tropical waters, preferring deep water – they are normally considered to be an offshore species. They have been known to mass strand, probably because of strong social affiliations within the pods. The largest such event occurred in 1946, when more than 800 individuals came ashore near Mar del Plata in Argentina, while in New Zealand 100 stranded on the Chatham Islands in 1906 and 253 washed up on the Waiau mudflats in Auckland's Manukau Harbour in 1978.

False killer whales have a low reproductive rate and nurse their calves for up to two years. It is thought that females become reproductively senescent at around 45 years of age. The whales are often aggressive towards other cetacean species and have been observed chasing smaller dolphins and even attacking larger whales. The species has mated with bottlenose dolphins in captivity, producing offspring known as 'wholpins'.

Pseudorca have been observed swimming in long, mile-wide formations, possibly reflecting the fact that they are cooperative hunters that work together to catch prey. Examination of their stomach contents has shown that they eat a wide variety of cephalopods and fish, including large species such as mahi-mahi, yellowfin tuna and swordfish. In fact, in some areas the whales are notorious for stealing fish weighing up to 30 kilograms from the hooks of long-line fishermen. In New Zealand they have been observed working cooperatively with bottlenose dolphins to herd and capture prey such as kahawai.

NEW ZEALAND DISTRIBUTION

In New Zealand, false killer whales are most commonly sighted along the east coast around the top half of the North Island, and around the Chatham Islands.

THREATS AND CONSERVATION

Pseudorca are deliberately killed in drive-and-harpoon fisheries off Japan, and are hunted opportunistically in Indonesia and the Caribbean. They are also killed in the long-line fishing industry, sometimes in direct shooting incidents and at other times as a by-catch. In addition, the whales are sometimes captured for display in marine parks. Although orca predation obviously occurs, it is not commonly observed and does not affect false killer whale population numbers. The whales live a relatively safe life in South Pacific and New Zealand waters, and are not considered under threat.

Pseudorca *(background) are highly gregarious and usually occur in large pods, often in the company of common bottlenose dolphins (foreground).*

New calves are pale grey in colour and are between 1.6 and 2 metres long.

Southern right whale dolphin

Lissodelphis peronii (Lacépède, 1804)

Size	At least 3 m, up to 116 kg
Gestation	Unknown
Birth size	Probably 1–1.1 m
Calving interval	Unknown
Lifespan	Unknown
Teeth	88–100 pairs
Diet	Fish, especially lanternfish, and squid
IUCN status	Data Deficient
New Zealand population	Resident year-round; occasionally sighted; Not Threatened

ETYMOLOGY

There are two right whale dolphin species, one in the northern hemisphere and the other in the southern hemisphere. The two species share the genus name *Lissodelphelis*, from the Greek *lisso*, meaning 'smooth' and referring to their smooth backs (a unique feature among dolphins), which lack a dorsal fin. The southern species carries the specific name *peronii* in recognition of the French naturalist François Peron, who first described the dolphins off the coast of Tasmania in 1802.

DESCRIPTION

Right whale dolphins were named after right whales because, like the whales, they lack a dorsal fin – in the case of southern right whale dolphins, they are the only dolphin species in the southern hemisphere without this structure. The species is extremely slender and sleek, with a black and white colour pattern (northern right whales are mostly black with white ventral patches).

BEHAVIOUR

The dolphins are a poorly known species owing to their offshore existence, so little is known of their population numbers or biology. They commonly associate with other dolphin species such as long-finned pilot whales and bottlenose and dusky dolphins. The dolphins feed on fish such as lanternfish as well as various squid species. Little is known about their feeding habits, but dive times of six to seven minutes have been recorded for foraging individuals.

NEW ZEALAND DISTRIBUTION

Southern right whale dolphins are observed most often in cool, deep, offshore waters and live mainly in the subantarctic between latitudes 40°S and 55°S. They are reasonably common off New Zealand's South Island and have occasionally been observed in the deep waters off Kaikoura. Southern right whale dolphins are known to strand on occasion, with events being recorded from Foveaux Strait, in the deep south, to Whananaki, in the far north. One of the largest recorded strandings of the species occurred in 1988, when at least 75 animals came ashore on the Chatham Islands.

THREATS AND CONSERVATION

Southern right whale dolphins are known to be caught in the swordfish gillnet industry off the coast of Chile, but little is known about the effects of other threats – including pollution – on the species. Their preference for living offshore makes them relatively unthreatened as fishing by-catch in New Zealand waters.

In his 1851 novel Moby-Dick, *Herman Melville described the southern right whale dolphin as having a 'neat and gentleman-like figure … But his mealy-mouth spoils him'.*

The dolphins have a unique swimming style and often travel in a series of long, low leaps, making them appear to bounce across the sea.

Risso's dolphin

Grampus griseus (Cuvier, 1812)

Size	4.0–4.3 m, up to 450 kg
Gestation	Unknown
Birth size	1.1–1.5 m
Calving interval	Unknown
Lifespan	Up to 35 years
Teeth	2–7 pairs of stout, pointed teeth in front of lower jaw and, on rare occasions, 1–2 vestigial pairs in upper jaw
Diet	Almost exclusively squid
IUCN status	Least Concern globally; Mediterranean sub-population Data Deficient
New Zealand population	Vagrant

ETYMOLOGY

The genus name *Grampus* comes from the Latin *grandis*, meaning 'large', while the species epithet, *griseus*, means 'mottled with grey'. The species was first described from a dolphin that had been found stranded on a beach in Brittany and then stuffed. The French naturalist Antoine Risso described it to zoologist Georges Cuvier, who gave it the common name Risso's dolphin in honour of his colleague. In 1812, Cuvier assigned the name *Delphinus griseus* to the dolphin, and in 1928 the genus name was changed to *Grampus*.

DESCRIPTION

Risso's are large, stocky dolphins with a tall, erect dorsal fin and a blunt, beakless head. Their broad, squarish melon or forehead is creased in front with a longitudinal groove, and their bodies tend to be heavily marked with linear and circular scars made by other Risso's dolphins and cookiecutter sharks.

The calves are light fawn or silver-grey when born, becoming black and then mottled grey as they mature; very old dolphins are sometimes almost pure white. The anchor-shaped ventral chest patch remains white and the dorsal fin dark grey throughout the dolphin's life.

BEHAVIOUR

Risso's are found throughout the world in temperate and tropical waters in oceans and large seas. They feed primarily at night, when their squid prey migrates towards the surface, and have been observed swimming abreast in an evenly spaced echelon formation, believed to be a hunting tactic. Risso's are gregarious and are often seen in groups of 25–50 animals, and sometimes in groups that number into the hundreds. They are observed with other species such as striped and common dolphins and pilot whales.

NEW ZEALAND DISTRIBUTION

Risso's dolphins are infrequently seen in New Zealand waters. The most famous individual was Pelorus Jack, a solo Risso's dolphin that lived in the Marlborough Sounds for at least 24 years and possibly longer. In January 1983, six male Risso's were found inside Whangarei Harbour.

Over a week-long period, four of the dolphins stranded and had to be rescued, while two others were successfully guided through the sandbanks to deep water. But by the end of the week, all six had returned to shallow waters and subsequently died. In March 2010, a Risso's was found stranded on the Chatham Islands, the first record there. A live Risso's sighting was also reported off the Kaipara coast in March 2001, although the identification was not confirmed.

THREATS AND CONSERVATION

Although there are no definitive population estimates, Risso's are believed to be common throughout their range. They continue to be hunted in Japan, where an estimated 250–500 are killed annually, and are taken on occasion in traditional fisheries in the Philippines, Sri Lanka and the Solomon Islands. They are also reported as by-catch in gillnet and trawl fisheries.

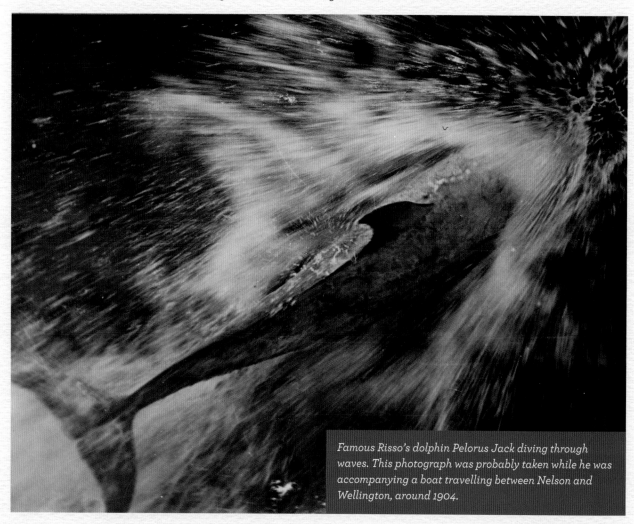

Famous Risso's dolphin Pelorus Jack diving through waves. This photograph was probably taken while he was accompanying a boat travelling between Nelson and Wellington, around 1904.

Melon-headed whale

Peponocephala electra (Gray, 1846)

Size	Up to 2.8 m, up to 275 kg
Gestation	12 months
Birth size	1.1 m, 15 kg
Calving interval	Unknown
Lifespan	Male at least 22 years; female at least 30 years
Teeth	40–52 pairs in both lower and upper jaws
Diet	Fish, squid and, occasionally, crustaceans
IUCN status	Least Concern
New Zealand population	Vagrant

ETYMOLOGY

The species was originally placed in the genus *Lagenorhynchus* but in 1966 was reassigned to its own genus, *Peponocephala*. The whale is also sometimes known as the electra dolphin, after its species epithet, meaning 'bright'.

DESCRIPTION

Melon-headed whales are mostly dark grey with a faint lighter grey dorsal cape and a pale band extending from the tip of the melon to the blowhole. Males are slightly larger than females and have a taller dorsal fin and larger tail flukes.

BEHAVIOUR

Melon-headed whales are a deep-water species ranging throughout tropical and semi-tropical oceans. They are very gregarious, usually occurring in pods of several hundred. Although squid appears to be their preferred prey species, small fish and shrimps have also been found in the stomachs of stranded individuals. Biological information gathered at strandings indicates that calving takes place between late winter and spring (August to December). Females are believed to reach sexual maturity at about 13–14 years of age and males at around 15 years.

NEW ZEALAND DISTRIBUTION

There has never been a sighting of a live individual in New Zealand waters. In January 2007 a single 2.4-metre melon-headed whale stranded at Mercury Bay and its complete skeleton was gifted to Te Papa by local iwi.

THREATS AND CONSERVATION

Pollution and occasional capture in gillnet fisheries in the Philippines, Indonesia and Malaysia are the main known threats these whales face.

Pygmy killer whale

Feresa attenuata (Gray, 1874)

Size	2.1–2.6 m, up to 225 kg
Gestation	Unknown
Birth size	80 cm
Calving interval	Unknown
Lifespan	Unknown
Teeth	20–26 pairs
Diet	Fish and squid
IUCN status	Data Deficient
New Zealand population	Vagrant

ETYMOLOGY

The pygmy killer whale's species name *attenuata* (Latin) means 'thinning' and derives from the fact that the whale's relatively robust body narrows towards the dorsal fin. Two skulls were the only evidence of the whale's existence until the 1950s when Japanese scientist Munesato Yamada examined a 'porpoise' that had been killed by whale and dolphin hunters at Taiji. He proposed that the dolphin be known as the 'Lesser' or 'Pygmy' killer whale. The species is considered an oceanic dolphin.

DESCRIPTION

The animal lacks a beak and its melon is rounded and extends slightly forward of the front of the mouth. The dorsal is tall and upright, while the flippers are moderately long and tapered. The body is dark with lighter grey on the dorsal sides and white around the lips and on the belly.

BEHAVIOUR

The pygmy killer whale has a worldwide tropical and temperate distribution. They are generally sighted in groups of fewer than 50 individuals. Fish and squid appear to be their favoured prey. Pygmy killer whales have also been heard making 'growling' noises at the surface.

NEW ZEALAND DISTRIBUTION

Pygmy killer whales are only known in New Zealand waters from a stranding of a dead animal on a Northland beach on 24 December 2010.

THREATS AND CONSERVATION

A few animals are taken in drives and driftnets, notably in Japan and Sri Lanka. Incidental catches are also reported in gillnet fisheries and tuna purse seines.

Fraser's dolphin

Lagenodelphis hosei (Fraser, 1956)

Size	Up to 2.7 m, up to 210 kg
Gestation	12–13 months
Birth size	1 m, 19 kg
Calving interval	Estimated at 2 years
Lifespan	At least 16 years, probably longer
Tooth	36–44 pairs in upper jaw, 34–44 pairs in lower jaw
Diet	Fish, squid and crustaceans
IUCN status	Least Concern
New Zealand population	Vagrant

ETYMOLOGY

The genus to which Fraser's dolphin belongs was proposed in the mid-1950s by Francis Fraser, a scientist at the British Museum who examined a skull found in 1895 in Sarawak, Borneo, and realised it was a new species. The dolphin's specific name was assigned in honour of Charles Hose, the British naturalist who originally collected the bone fragments. Fraser's dolphins were not seen in the flesh until 1970, when a calf was caught in a tuna purse seine net off California.

DESCRIPTION

Fraser's dolphins possess a very short beak and small flippers, and their dorsal fin is small and slightly hooked. Their most striking feature is a bold, dark lateral stripe around their eyes.

BEHAVIOUR

Little is known about the behaviour, life history or population numbers of this species. The dolphins are usually sighted in pods of around 100 or more.

NEW ZEALAND DISTRIBUTION

A pod of dolphins was photographed in November 2004 from a Royal New Zealand Air Force plane 240 kilometres northeast of North Cape. Cetacean expert Alan Baker later identified the dolphins as Fraser's, in the only confirmed live sighting in New Zealand waters to date.

THREATS AND CONSERVATION

Fraser's dolphins are susceptible to being caught in tuna purse seine nets and have been captured in driftnets and anti-shark barrier nets. They were hunted in the South Pacific both as a source of food and for their teeth. Ocean pollution is undoubtedly one of their biggest threats today.

Pantropical spotted dolphin

Stenella attenuata (Gray, 1846)

Size	2.2–2.5 m, 120 kg
Gestation	11 months
Birth size	85 cm, weight unknown
Calving interval	2–4 years
Lifespan	40–45 years
Teeth	35–48 pairs in upper jaw, 34–47 pairs in lower jaw
Diet	Small fish, squid and crustaceans
IUCN status	Least Concern
New Zealand population	Vagrant

ETYMOLOGY

Both the genus and species names are Latin for 'thin', a reference to the dolphin's long, slender beak.

DESCRIPTION

As young calves mature, spots start to appear on their skin. The spotting and striping patterns are complex, and vary between individuals and regions.

BEHAVIOUR

Pantropical spotted dolphins are one of the most abundant cetacean species, numbering an estimated 3 million. The dolphins are extremely gregarious and swim in schools of hundreds and even thousands. The dolphins often engage in aerial acrobatics and bow-riding, and they are fast – two trained pantropical spotted dolphins reached speeds of 39 kilometres per hour in just two seconds!

NEW ZEALAND DISTRIBUTION

In New Zealand, a skull of the species was found in Golden Bay in the South Island in 1869. A large school of the dolphins was identified between East Cape and the Kermadec Islands in 1975, indicating that the species does stray into New Zealand waters on occasion.

THREATS AND CONSERVATION

By the early 1990s, the annual kill of pantropical spotted dolphins in tuna purse seine nets had been reduced to around 15,000 animals, and since then it has been reduced further still. The dolphins have also been killed by drive fisheries in Japan and the Solomon Islands, and have been taken in Sri Lanka, the Caribbean and the Philippines, where they are killed for food and bait.

Striped dolphin

Stenella coeruleoalba (Meyen, 1833)

Size	Up to 2.65 m, up to 156 kg
Gestation	12–13 months
Birth size	93–100 cm
Calving interval	Unknown
Lifespan	Up to 58 years
Teeth	80–110 pairs
Diet	Small fish, lanternfish, cod and squid are common
IUCN status	Least Concern
New Zealand population	Vagrant

ETYMOLOGY

The striped dolphin's species name, *coeruleoalba*, is derived from the Latin *caeruleus*, meaning 'sky blue', and *albus*, meaning 'white', and refers to the animal's striking bluish-grey and white colour pattern.

DESCRIPTION

These dolphins have stripes that run from behind the eye to the mid-belly, from behind the eye to the back edge of the flipper, and additional stripes that run from in front of the eye to the front edge of the flipper.

BEHAVIOUR

The dolphins are found worldwide in tropical and temperate waters. They often travel in dense pods containing from 25 up to several hundred individuals. The dolphins exhibit a unique behaviour known as roto-tailing, where they make high arcing leaps while rapidly rotating their tails. The dolphins feed on a variety of pelagic and benthopelagic fish and squid. It appears that females become sexually mature between five and 13 years, and males between seven and 15 years.

NEW ZEALAND DISTRIBUTION

A single dolphin was found stranded at Waikanae Beach near Wellington in 1870. In 1971, a single striped dolphin washed ashore at Auckland's Mission Bay. There are still few confirmed live sightings in New Zealand waters.

THREATS AND CONSERVATION

Striped dolphins were targeted for many years in the Japanese dolphin drive hunt and have suffered significant losses in the Mediterranean in the pelagic driftnet fishery. Overfishing and pollution are also taking their toll on the population.

Rough-toothed dolphin

Steno bredanensis (Cuvier, 1828)

Size	Up to 2.65 m, up to 160 kg
Gestation	Unknown
Birth size	1 m, weight unknown
Calving interval	Unknown
Lifespan	At least 32 years
Teeth	19–26 pairs in upper jaw, 19–28 pairs in lower jaw
Diet	Fish, squid and octopus
IUCN status	Least Concern
New Zealand population	Vagrant

ETYMOLOGY

Steno comes from the Greek word for 'narrow' and refers to the dolphin's distinctive beak, which is not separated from its forehead by a crease, as with other long-beaked dolphins. The species name acknowledges Pieter van Breda, an artist who first noticed the species in Georges Cuvier's notes and descriptions. The dolphin's common name refers to the ridges caused by thin lines of enamel that run vertically down each of its teeth.

DESCRIPTION

The dolphin's dorsal fin and flippers are relatively large. The tip of its slender beak is generally white. White spots are usually found on the body, possibly scars from the bites of cookiecutter sharks.

BEHAVIOUR

Rough-toothed dolphins are widely distributed in deep tropical and warm temperate waters. They are usually found in close-knit groups averaging 10–30 individuals.

Little is known about the biology or lifestyle of these dolphins. They feed on squid and fish. They are unpopular with fishermen in Hawai'i, where they are known to steal both bait and fish from fishing lines.

NEW ZEALAND DISTRIBUTION

In New Zealand, one stranded specimen was found in the Bay of Plenty in 1990, and, on 18 June 1990, four stranded at Woodside Creek south of Blenheim, their most southerly New Zealand record.

THREATS AND CONSERVATION

Rough-toothed dolphins are not seriously threatened by human activity within their range.

PHOCOENIDAE: PORPOISES

Porpoises differ from dolphins in that they generally have shorter beaks, tend to be smaller and they have spade-shaped rather than conical teeth. There are seven porpoise species worldwide. Only the little-known spectacled porpoise has been recorded in the southwest Pacific.

A female spectacled porpoise.

Spectacled porpoise

Phocoena dioptrica (Lahille, 1912)

Size	Up to 2.3 m, up to 115 kg
Gestation	Unknown
Birth size	90–100 cm, weight unknown
Calving interval	Unknown
Lifespan	Unknown
Teeth	Spade-shaped; 17–24 pairs in upper jaw, 16–20 pairs in lower jaw
Diet	Anchovies and crustaceans have been found in stranded specimens
IUCN status	Data Deficient
New Zealand population	Vagrant

ETYMOLOGY

This little-known cetacean was first described by the Argentinian naturalist Lahille, who gave it the species name *dioptrica*, from the Latin word for 'spectacled' – the porpoise has a fine white line around the top of each eye that resembles eye glasses.

DESCRIPTION

Before the mid-1970s, spectacled porpoises were known from only nine stranded specimens and one sighting. Even today, the majority of information on the species has come from remains found on remote beaches. The porpoises have a jet-black upperside and a white underside. Males have a large, almost rounded, dorsal fin, while the female's dorsal is smaller and more triangular (detail above). Like all porpoise species, they have robust bodies, a bluntly pointed snout with no beak, and small spade-shaped teeth.

BEHAVIOUR

The diet and feeding behaviour of the spectacled porpoise are largely unknown, although anchovies and small crustaceans have been found in the stomachs of stranded individuals. The few glimpses of live animals indicate that they occur singularly or in groups of two to three individuals. They appear to live in both coastal and offshore waters, have a circumpolar range and are found in both subantarctic and Antarctic waters. Most of the sightings and stranding records are from the Tierra del Fuego coast at the southern tip of South America. In Australasian waters, the porpoises have been sighted around Tasmania and Macquarie Island.

NEW ZEALAND DISTRIBUTION

The first documentation of the spectacled porpoise's presence in New Zealand occurred in 1975, when a skull was collected from the Auckland Islands. Since then, there have been

Male spectacled porpoises have a large, almost rounded dorsal fin.

live sightings at both the Auckland and Antipodes islands, and strandings in 1999 near Nugget Point on the South Island, 2006 at Motueka at the top of the South Island, and 2011 on the Coromandel Peninsula.

THREATS AND CONSERVATION

Based on the low number of sightings, spectacled porpoises appear to be rare, but their total population is currently unknown. There are records of the porpoise being shot and eaten by 20th-century whalers near South Georgia, and fishermen off the coast of Uruguay traditionally hunted them for food. It is believed that some (and possibly most) of the porpoises washed up on beaches in Tierra del Fuego were incidental by-catch in the coastal gillnet fishery. However very little is known about these 'spectacled' creatures.

ZIPHIIDAE: BEAKED WHALES

Beaked whales, known to Māori as hakura, are the strangest, the most mysterious and the least understood of all cetaceans. Ziphiidae species are rarely observed in the wild because they are deep-diving, open-ocean animals with generally cryptic behaviour. Scientists currently recognise 21 different species, some of which have never been seen alive and are known to exist only from dead animals that have washed ashore or been caught by fishermen. In recent years, however, genetic analysis has begun to reveal new information about some of these enigmatic whales. In 2002, a new species, Perrin's beaked whale, was recognised based on DNA analysis of five animals that stranded along the California coast, and in 2012 genetic testing on the skeletons of two whales that stranded on a beach in New Zealand revealed that the whales were spade-toothed whales, one of the rarest of all the beaked whale species.

Beaked whales range in size from just under 4 metres to nearly 13 metres long. Most of them have a fairly pronounced beak, a pair of throat grooves in the shape of a 'V', and slight depressions along their sides called flipper pockets, where they can tuck their relatively small flippers away. But the most remarkable characteristic of all is their teeth – or, rather, their lack of them. The vast majority of female and immature beaked whales have no erupted teeth, while the males have some of the weirdest teeth in the animal kingdom. Most males have between two and four functional teeth, which are found only in their lower jaw. In some animals, the teeth erupt outside the mouth and are more like tusks.

Most beaked whales are deep divers and are usually found in waters that are at least 300 metres deep. Based on stomach analysis of stranded animals, it appears the majority of their prey is mesopelagic or bathypelagic squid and fish species, which are sucked in by the feeding whales. Like sperm whales, these deep-diving cetaceans appear to spend much of their time foraging. Their elusive behaviour at the surface makes them extremely difficult to identify at sea, so behavioural studies on most species is very limited, as are any estimates of population size.

Of the 21 known beaked whale species, 13 have been identified in New Zealand waters. Because so little is known about these whales, the fact files given for each species listed here are brief compared to those for other groups. For the same reason, sections in this book on the species' behaviour and on threats and conservation are limited or have been omitted. Like all other cetacean species, beaked whales are threatened by environmental factors such as global warming and pollution.

The distinctive head of a Blainville's beaked whale.

Cuvier's beaked whale

Ziphius cavirostris (Cuvier, 1823)

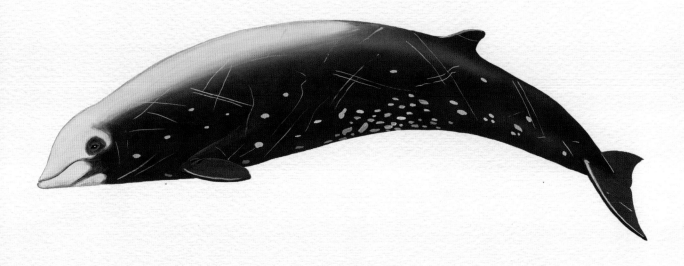

Size	Up to 7 m, up to 3000 kg
IUCN status	Least Concern globally; Mediterranean sub-population Data Deficient

ETYMOLOGY

Cuvier's beaked whale is named after French scientist Georges Cuvier, who first described the species in 1823. *Cavirostris* is Latin and means 'hollowed or concave beak' and refers to a diagnostic hollow or concavity in the animal's skull. The whales are also commonly referred to as goose-beaked whales.

DESCRIPTION

Cuvier's beaked whales have a rotund body shape and are dark grey to reddish brown in colour. The shape of the head and short beak is distinctive and said to resemble the head of a goose. Males have a white head. The males have two forward-pointing conical teeth erupting from the tip of the lower jaw. The teeth are visible outside the mouth and are sometimes covered with barnacles.

BEHAVIOUR

Cuvier's are normally found in waters that range from depths of 1500–3500 metres. Based on analysis of stomach contents, the species' main prey is deep-water squid.

NEW ZEALAND DISTRIBUTION

The species is relatively common in New Zealand waters, being known from live sightings and 80 or so strandings throughout the country, from the Kermadecs to the bottom of the South Island.

THREATS AND CONSERVATION

It is thought that these whales are particularly susceptible to acoustic trauma – mass strandings in both the Bahamas and the Mediterranean Sea have been linked to military exercises involving powerful underwater sound transmissions.

Southern bottlenose whale

Hyperoodon planifrons (Flower, 1882)

Size	6.9–7.5 m, weight unknown
IUCN status	Least Concern

ETYMOLOGY

The genus name is actually a misnomer and refers to what were mistakenly considered to be small teeth but are in fact merely bony bumps on the whale's palate. The species epithet, *planifrons*, means 'flat-fronted' and refers to the whale's distinctive forehead – an alternative common name is flat-headed whale.

DESCRIPTION

Southern bottlenose whales are greyish brown to dull yellow in colour and their body is often scarred with scratches and light splotches. Males are often very pale and appear almost white at sea. Their head is distinctive, with a steep forehead, bulbous melon and a well-defined short, stubby beak. Males have a pair of small conical teeth that erupt at the tip of the lower jaw.

BEHAVIOUR

Southern bottlenose whales are circumpolar in the southern hemisphere, ranging from cold, deep waters in Antarctica to about 30°S. The species is the most common beaked whale seen in Antarctica and in summer animals are often observed within 100 kilometres of the ice edge. They can remain underwater for over an hour and feed on deep-water squid, as well as fish species such as the Patagonian toothfish.

NEW ZEALAND DISTRIBUTION

More than 20 southern bottlenose whales are reported to have stranded in Australia and New Zealand. One that stranded in 1979 at Ohope in the Bay of Plenty had more than 200 squid beaks in its stomach.

Shepherd's beaked whale

Tasmacetus shepherdi (Oliver, 1937)

Size	Up to 7 m, weight unknown
IUCN status	Data Deficient

ETYMOLOGY

The species was first described by W.B.B. Oliver, naturalist and director of the Dominion Museum (predecessor of Te Papa) from a stranded individual collected on the Taranaki coast in 1937 by George Shepherd, curator at Wanganui's Alexander Museum. The genus name refers to the fact that the whale specimen came from the Tasman Sea.

DESCRIPTION

Shepherd's beaked whales are the only members of the Ziphiidae family with a full set of functional teeth. The upper body is dark with patches of white extending upwards from the belly. The beak is long and slender, and the melon is prominent and also displays a white patch.

BEHAVIOUR

Today we know little more than we did in 1937. There have been only a few reported live sightings, none of which has been confirmed, and most known information has come from stranded specimens.

NEW ZEALAND DISTRIBUTION

Around 20 strandings of Shepherd's beaked whales have been recorded in New Zealand, most of them from the south. The first recorded specimen came ashore at Wanganui in 1933.

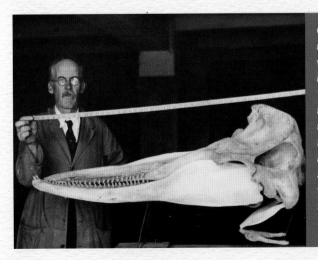

George Shepherd measures the whale skull which he found on a Taranaki beach in 1937. The skull turned out to be from an unknown beaked whale species which was subsequently named 'Shepherd's' in his honour.

Arnoux's beaked whale

Berardius arnuxii (Duvernoy, 1851)

Size	Up to 9.3 m, weight unknown
IUCN status	Data Deficient

ETYMOLOGY

Arnoux's beaked whales were named by zoologist Georges Duvernoy after the French surgeon Dr Maurice Arnoux, who found a skull of the whale near Akaroa and presented it to the Paris Museum of Natural History in 1846. The genus name honours Admiral Auguste Bérard, captain of the ship that brought the specimen to France.

DESCRIPTION

Arnoux's beaked whales are relatively long and slender, with a dark grey to light brown body that displays numerous white linear scars and scratches. They have a long dolphin-like beak and a bulbous melon. They have two pairs of triangular teeth at the tip of the lower jaw; the forward pair are larger and visible outside the closed mouth, while the smaller pair are concealed inside the mouth.

BEHAVIOUR

The whales have a circumpolar distribution in the southern hemisphere, with most sightings recorded south of 40°S. They are shy and elusive, and consequently little is known of their biology, although they are believed to be deep divers. Although the whales are often sighted in small groups of 10 or less, researchers in the Antarctic tracked one large group of about 80 whales for several hours until they dispersed into the loose pack ice. Most of the dives timed by the researchers lasted 15–25 minutes, but one dive lasted over an hour, with the whales travelling almost 6 kilometres before resurfacing.

NEW ZEALAND DISTRIBUTION

About 50 strandings have been recorded in New Zealand. Arnoux's are not exploited and there is no indication that their populations are in any immediate danger.

Gray's beaked whale

Mesoplodon grayi (von Haast, 1876)

Size	Up to 5.6 m, at least 1100 kg
IUCN status	Data Deficient

ETYMOLOGY

Roughly translated, the Latin genus name *Mesoplodon* means 'armed with a tooth in the middle of the jaw'. Gray's beaked whale was described in 1876 and named after John Edward Gray, a zoologist at the British Museum. The whale is sometimes referred to as the Scamperdown whale.

DESCRIPTION

Gray's are unique among members of the *Mesoplodon* genus in that both sexes have teeth. The body of Gray's is slender with a dark grey upperside and pale underside. Adults have a long, slim white beak, which is often lifted out of the water at a 45° angle.

BEHAVIOUR

Gray's are primarily a southern hemisphere species, with most strandings occurring between 30°S and 45°S, and most sightings in subantarctic waters, although some are also known from the Antarctic. The northern coast of the North Island and the Chatham Islands are stranding hotspots for the species. An increase of females stranding with dependent calves between December and May suggests that females may move closer to shore during their summer calving season. Based on the number of group strandings, it is thought these whales may associate in larger aggregations than those normally reported for other beaked whale species.

NEW ZEALAND DISTRIBUTION

There are more than 180 recorded strandings of Gray's beaked whales in New Zealand, which ranks them among the most common stranders in these waters, along with pygmy sperm and long-finned pilot whales. A mass stranding of 28 individuals occurred on the Chatham Islands in 1874, while more recently, in January 2012, four Gray's beaked whales beached at Papamoa in the Bay of Plenty.

Hector's beaked whale

Mesoplodon hectori (Gray, 1871)

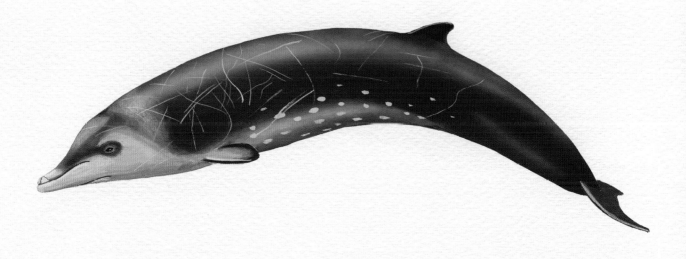

Size	Up to 4.5 m, weight unknown
IUCN status	Data Deficient

ETYMOLOGY

The Hector's beaked whale, like the Hector's dolphin (see page 198), was named after Sir James Hector, director of the Colonial Museum (predecessor to Te Papa).

DESCRIPTION

Very few live Hector's beaked whales have ever been examined, so the species' body colour was virtually unknown until recently. In 2002, scientists were able to examine two fresh specimens (one male and one female) that stranded live off the coast of Buenos Aires in Argentina. The male had a dark greyish-brown back, shading to a lighter colour, and a pale white beak with a single pair of flattened triangular teeth near the tip of the lower jaw. In contrast, the female had a light grey back with a pale white belly. In 2002, scientist Nick Gales described a live sighting in shallow waters off Western Australia of a dark brown juvenile beaked whale with a light lower jaw; this was later identified as a Hector's beaked whale using DNA analysis.

BEHAVIOUR

Aside from specimens stranded in New Zealand, the species has also been identified from strandings in South America, South Africa and Tasmania. Hector's are thought to occur in the southern hemisphere and are considered to be one of the rarer beaked whales.

NEW ZEALAND DISTRIBUTION

As of 2010, only around 40 stranded specimens had been positively identified as Hector's beaked whales, with the largest sample – 16 specimens – being found in New Zealand waters. The majority of strandings occur between December and April, suggesting that the whales may move into inshore waters during the summer.

Strap-toothed whale

Mesoplodon layardii (Gray, 1865)

Size	Up to 6.2 m, 1500 kg
IUCN status	Data Deficient

ETYMOLOGY

Strap-toothed whales were described in 1865 by British taxonomist John Gray from drawings sent by Edgar Layard, curator at the South African Museum, after whom the species is named. The species is also referred to as Layard's beaked whale.

DESCRIPTION

Strap-toothed whales are the largest *Mesoplodon*. Adults have a distinctive dark and light colour pattern that makes them relatively easy to identify at sea: the long beak and throat area is usually greyish white, while the top of the head and the area around the eye remain dark. The light and dark colours on calves are reversed, changing to adult colouring as the calf matures. The species' most distinctive feature, however, and the one that gives it its common name, is the male's two bizarre teeth. These emerge from near the middle of the lower jaw and then curl upwards and backwards at about 45° until they extend over the upper jaw (detail above). The teeth, or tusks, are often covered with stalked barnacles, are up to 30 centimetres long and sometimes prevent the whales from opening their jaws wide. It is believed they use a vacuum-cleaning system of feeding, sucking in mouthfuls of relatively small squid species.

BEHAVIOUR

Strap-toothed whales are hard to approach at sea but have been observed from a distance basking on the calm surface of the water. They appear not to raise their tails when diving, sinking slowly or doing a sideways roll to slip under the water. Dives appear to last 10–15 minutes.

The whales appear to be distributed throughout the cold temperate oceans of the southern hemisphere, with most sightings occurring between latitudes 33°S and 53°S.

NEW ZEALAND DISTRIBUTION

This species strands commonly in New Zealand, with most events taking place in the northern half of the North Island between January and April, indicating that the whales move closer inshore during the summer months.

Andrews' beaked whale

Mesoplodon bowdoini (Andrews, 1908)

Size	Up to 4.5 m, weight unknown
IUCN status	Data Deficient

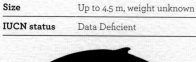

ETYMOLOGY

The Andrews' beaked whale species names honours George S. Bowdoin, a trustee and donor to the American Museum of Natural History.

DESCRIPTION

Andrews' beaked whales are a dark bluish-black colour all over and possess a low melon. The front half of the whale's short, wide beak is white, and in males a large, flat tooth erupts from each side of the arched lower jaw. The exposed tips of these teeth protrude slightly above the upper jaw and are sometimes covered with barnacles.

BEHAVIOUR

This is another poorly known beaked whale, identified from a few specimens found stranded between 32°S and 55°S, north of the Antarctic Convergence. It had been suggested that Andrews' beaked whale might be conspecific with the northern hemisphere Hubb's beaked whale, but in 2001 New Zealand scientist Alan Baker published a paper that confirmed Andrews' as a distinct species that could be distinguished both genetically and by morphological features such as the shape of its teeth and the differences in its skull. Body scars indicate fighting between males.

NEW ZEALAND DISTRIBUTION

The first specimen described was an individual that stranded on New Brighton Beach, Canterbury, in 1904. At least 10 other New Zealand strandings have occurred, from Northland's Whangarei Heads to the subantarctic Campbell Island.

The examination of foetuses present in females that stranded in May and September, and the presence of newly born calves with females that stranded in September, indicate a spring breeding season in New Zealand, with the length of new calves estimated to be around 2.2 metres.

Blainville's beaked whale

Mesoplodon densirostris (Blainville, 1817)

Size	Up to 4.7 m, up to 1030 kg
IUCN status	Data Deficient

ETYMOLOGY

One of the first beaked whale species was discovered in 1817, when French scientist Henri de Blainville examined a piece of jawbone that he described as being the 'heaviest bone that he had ever seen, denser even than elephant ivory'. His description led to the identification of a new species known as Blainville's, or dense-beaked, whale, whose species epithet, *densirostris*, means 'dense-beaked'.

DESCRIPTION

Blainville's is a distinctive beaked whale species. The lower jaw is highly arched in both sexes, but in adult males the arches are especially extreme, with two large flattened teeth erupting from the crest of the jaw at a 45° angle. The tips of the male's teeth extend above the upper jaw and are often covered with barnacles. The beak is thick and moderately long, while the forehead appears flattened.

The whales have a brownish or bluish-grey colour on top with pale undersides. The bodies of males often display round or oval white scars from cookiecutter shark bites, along with widely separated, paired scratches, the latter presumably caused by the teeth of other males.

BEHAVIOUR

Blainville's beaked whales have the most extensive known distribution of any *Mesoplodon*. They inhabit tropical and temperate oceans worldwide, and are observed on a regular basis in the Bahamas, Hawai'i and the Society Islands of the South Pacific.

This species is one of the most widely studied of the beaked whales. It is normally sighted in waters that are at least 500–1000 metres deep, often with deeper gullies nearby. Dives lasting 54 minutes and reaching depths up to 1400 metres have been recorded. Adult males appear to have small harems of females, while sub-adults are usually sighted in separate groups. A stranding of several whales in the Bahamas in 2000 was linked to acoustic trauma from military exercises that had taken place just prior to the event.

NEW ZEALAND DISTRIBUTION

The whales are known in New Zealand from strandings in the north of the country.

Ginkgo-toothed beaked whale

Mesoplodon ginkgodens (Nishiwaki and Kamiya, 1958)

Size	Up to 5.3 m, weight unknown
IUCN status	Data Deficient

ETYMOLOGY

Ginkgo-toothed beaked whales were first found and described in Japan in 1958. The teeth of males are shaped like the leaves of the ginkgo tree, hence the common and scientific species names.

DESCRIPTION

The colour pattern of these whales is not well known, since very few fresh specimens have been examined. Males appear to be dark grey with lighter undersides, while females appear to be lighter in colour.
The two wide, flattened teeth of adult males barely erupt from small arches slightly behind the middle of the lower jaw.

BEHAVIOUR

The little information that scientists have is mainly from a few dozen scattered strandings, most of which have occurred in temperate and tropical waters of the Indo-Pacific.

NEW ZEALAND DISTRIBUTION

The species is known in New Zealand from two strandings, one in Taranaki in 2003 and another in Golden Bay in 2004.

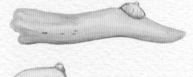

The gingko-leaf shaped tooth of a male, as it sits in the jaw (top) and in detail (bottom).

Pygmy beaked whale

Mesoplodon peruvianus (Reyes, Mead and Van Waerebeek, 1991)

Size	Up to 3.9 m, weight unknown
IUCN status	Data Deficient

ETYMOLOGY

Pygmy beaked whales are also known as Peruvian beaked or lesser-beaked whales, and have a scientific species name that originates with the story of their discovery. This began when a skull was found in a Peruvian fish market in 1976, which was later compared to the skull of a complete female specimen found in another fish market south of Lima in 1985. The first adult male was also found in Peru almost three years later. In fact, of the first 10 specimens of the species examined by scientists, at least six came from gillnets set for sharks off the coast of Peru. The first specimens to be discovered outside Peru occurred when two whales stranded in 1990 off La Paz in Baja, Mexico, followed by a December 1991 stranding near Kaikoura, New Zealand. The new species was officially named in 1991.

DESCRIPTION

Pygmy beaked whales are the smallest member of the family and have dark upper bodies with pale undersides. Their two barely visible teeth erupt from just in front of their slightly arched lower jaw and their beak is short and stubby. Their bodies display little visible scarring.

BEHAVIOUR

The species is known mainly from strandings, sightings or catches by fishermen off the coasts of Peru and Chile, but its range is obviously more extensive, as strandings have also occurred in Mexico, southern California and New Zealand. Almost all sightings have been of pairs, with one recorded sighting of two adults and one calf. Their behaviour is unknown as few live sightings have been confirmed. The whales appear to feed in mid to deep waters. The stomach contents of one stranded individual contained only fish, although another stranded specimen had a mix of fish and squid remains in its stomach.

NEW ZEALAND DISTRIBUTION

The Kaikoura stranding, in 1991, was of a 3.2-metre adult male. There is also an unconfirmed record of a pygmy beaked whale stranding in Golden Bay in 2011; the whale was subsequently refloated and then stranded again. It was refloated once more and not found again, so hopefully survived.

Spade-toothed whale

Mesoplodon traversii (Gray, 1874)

Size	Unknown (thought to be 4.5–5.5 m)
IUCN status	Data Deficient

ETYMOLOGY

A long and rather twisted tale led to the recognition of the beaked whale species known as the spade-toothed whale. In 1986, a partial skull was discovered on Chile's Robinson Crusoe Island, and in 1995 scientists declared it was a new beaked whale species, *Mesoplodon bahamondi*, with the common name Bahamonde whale. A short time later, New Zealand scientists realised that the skull from Chile was, in fact, genetically identical to another partial skull that had been recovered from White Island in the 1950s. Things became more interesting when they then extracted DNA from a jawbone and partial tusks from a specimen collected by Henry Travers at Pitt Island in the early 1870s. This whale was originally called *Mesoplodon traversii* after the collector and given the common name spade-toothed whale, but in 1878 James Hector declared it to be a strap-toothed whale, *Mesoplodon*

layardii (see page 241), and so the name *M. traversii* disappeared from the scientific literature. When the Pitt Island bone fragments were found to be genetically indistinguishable from the skull fragments from White Island and Chile, however, the scientific name *Mesoplodon traversii* and common name spade-toothed whale were resurrected.

DESCRIPTION

In December 2010, a beaked whale mother–calf pair stranded east of Opotiki in the North Island. The whales, originally identified as Gray's beaked whales (see page 239), were buried, but when DNA from tissue samples was later analysed, scientists from Auckland University were astounded and delighted to discover that the animals were in fact the rare and never before viewed spade-toothed whales. The adult female measured 5.3 metres, while her male calf was 3.5 metres long. Following

One of the first photos of a spade-toothed whale. This juvenile male stranded in 2010.

the discovery, the skeletons were exhumed, given a proper tangihanga (funeral) at Opape Marae and then taken to Te Papa for further morphological analysis.

The discovery of two complete spade-toothed whale skeletons around 140 years after bone fragments from the species were first found was particularly gratifying to Anton van Helden and Scott Baker, the scientists who had co-authored an earlier paper that described the Chile, White Island and Pitt Island specimens as belonging to the spade-toothed species.

BEHAVIOUR

While we still know little about the behaviour of the species, it can be surmised that the whales live mainly offshore in the South Pacific, and, like other beaked whales, they probably forage for deep-water prey, spending little time at the surface.

NEW ZEALAND DISTRIBUTION

The species is known in New Zealand from the Pitt Island jawbone, the White Island skull fragment and the skeletal remains of the mother and calf that stranded off Opotiki in 2010.

True's beaked whale

Mesoplodon mirus (True, 1913)

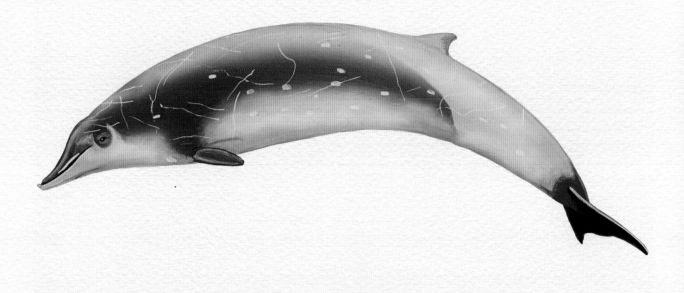

Size	5.3 m, up to 1400 kg
IUCN status	Data Deficient

ETYMOLOGY

The whale's common name refers to Fredrick W. True, who first described a specimen that had stranded on the coast of North Carolina in 1913. *Mirus* is Latin for 'extraordinary' or 'astonishing', referring to the find of the new species.

DESCRIPTION

True's beaked whales are rather rotund in the middle and taper off at both ends. In the North Atlantic, their body is generally medium or slightly brownish grey on the back and paler ventrally. The tip of the rostrum and dorsal fin are darker, and there is also a dark oval area around each eye. Southern hemisphere whales appear to be mostly bluish black, with a light dorsal fin and a light grey jaw. The melon is well rounded – almost bulbous – and slopes steeply to the short beak, and there is sometimes a deep crease in the area behind the blowhole. The male's teeth erupt slightly at the tip of the lower jaw, which extends just beyond the upper jaw.

BEHAVIOUR

Little is known about the whale's behaviour as live sightings are rare. Animals have been briefly observed in deep water (around 1000 metres), their blow is almost indistinct and they appear to surface for around 10–12 seconds before slipping under the water. It is believed that, like most beaked whales, they feed on deep-water squid and possibly fish species.

NEW ZEALAND DISTRIBUTION

On 25 November 2011, a whale stranded on rocks at Jackson Bay. Local fishermen refloated the whale, but it washed ashore on the Waiatoto Spit two days later. Tissue samples were taken and sent to Auckland University, and shortly after Christmas the whale was confirmed as a True's beaked whale, a species never before known in New Zealand waters.

MYSTICETI: BALEEN WHALES

Worldwide, there are no fewer than 14 Mysticeti (baleen) species. Eight species and two subspecies have been sighted in New Zealand waters, ranging from the largest – the mighty blue – to the smallest – the pygmy right. The southern right whale and pygmy right whale are each classed in separate families (Balaenidae and Neobalaenidae (or Cetotheriidae), respectively). The remaining six species and two subspecies are classed together in the family Balaenopteridae.

BALAENIDAE: RIGHT WHALES

The Balaenidae family includes the bowhead whale, along with three right whale species: the North Atlantic, the North Pacific, and the only southern hemisphere species, the southern right whale. Balaenids have large heads that are up to a third of their body length; long, narrow baleen; a highly arched upper jaw and huge bowed lower lips. The whales are large, rotund animals lacking a dorsal fin. Balaenids are skim feeders, swimming through large swarms of euphausiids (krill) or copepods with their mouths agape, allowing the water to escape while their prey becomes entangled in their long, very fine baleen. Bowhead whales are also known to feed on small schooling fish species. All Balaenidae species were heavily targeted by the whaling industry and suffered catastrophic losses; today, the North Pacific and North Atlantic right whales are considered to be the most endangered large whale species.

Diving with a right whale.

Southern right whale, Tohorā

Eubalaena australis (Desmoulins, 1822)

Size	Up to 17 m, up to 100,000 kg
Gestation	12 months
Birth size	4–4.6 m, 910 kg
Calving interval	3–5 years
Lifespan	At least 70 years
Baleen	Up to 2.6 m long, 220–270 per side; colour ranges from dark brown to grey or black
Diet	Copepods, occasional krill
IUCN status	Least Concern globally; Chile/Peru sub-population Critically Endangered
New Zealand population	Resident; estimated at 2300; Nationally Endangered

ETYMOLOGY

The southern right whale's genus name, *Eubalaena*, comes from the Greek *eu*, meaning 'true', and the Latin *balaena*, meaning 'whale'. Its specific name, *australis*, is Latin for 'southern'. To Māori, the whales are known as tohorā. The species' English common name comes from early whalers, who considered these whales to be the 'right' whales to catch: their massive bodies contain large amounts of oil, they are slow-moving and come close to shore, and they float when harpooned. Right whales were also hunted for their baleen, which is long (up to 2.6 metres) and plentiful (220–270 or more pieces hang from each side of the upper jaw). Worldwide, there are two other recognised right whale species: the North Atlantic right whale, *E. glacialis* (population in the hundreds), and the North Pacific right whale, *E. japonica* (population less than 100).

DESCRIPTION

Southern right whales are big – the largest recorded individual measured over 18 metres and weighed more than 115 tonnes. While the whales' average length of 15–17 metres is only half that of a blue whale, they weigh far more per metre of body length and can reach up to 100 tonnes, which gives them a rather rotund appearance. Their wide, dark grey to black body lacks a dorsal fin, their large flippers are up to 1.7 metres long and their tail flukes are 5–6 metres wide.

The whale's head is up to a third of its body length and is covered with callosities, raised patches of rough, hardened skin that are infested with at least three different species of cyamids (whale lice) that usually give them a greyish-white appearance. One southern right whale observed off Campbell Island had yellowish and pinkish callosities, and was consequently name Rock Garden. The callosities of each whale are uniquely arranged, and researchers use these patterns to differentiate between individuals. Interestingly, many of the callosities appear in places where you would find hair on a human – the top of the head, chin, upper and lower lips, and above the eyes. The largest callosity, found on the top of the head, is known as the bonnet.

BEHAVIOUR

The whales often lie on their sides, waving their paddle-shaped flippers in the air, and some behave like a yacht, raising their broad tail flukes high into the air in order to catch the wind and 'sail' along the surface of the water. Southern right whales are also quite acrobatic, and are often observed breaching, spyhopping and lob-tailing.

Right whales feed using a skimming technique. They swim through patches of krill or copepods with an open mouth, which allows the water to flow out the side as their tiny prey becomes entangled in the fine, hairy ends of their long baleen. They feed both under and along the surface of the water. One of their favoured prey, copepods, is also consumed by the whale lice living on their callosities. A skim-feeding whale may catch 2–3 tonnes of copepods in a day.

Some southern right whales, including the Australian, South African and Argentinean populations, migrate to warm waters at around 30°S to calve. Prior to the whaling era, New Zealand right whales gave birth in a number of sheltered bays around the mainland, but today the remnant population mainly mates and calves during winter in cold subantarctic waters off the Auckland and Campbell islands. During mating, a single female will swim along surrounded by an entourage of eager males, which jostle and push in their attempts to mate with her. A number of males will mate with a single female one after the other.

A male possesses testes weighing up to 500 kilograms each – they hold a lot of sperm, indicating that the mating game is essentially a sperm competition, with the last suitor often being the winner as his semen washes out that of his previous competitor. When the female becomes tired of all the attention, she simply turns belly up. Southern right whales have been observed mating year-round, but it appears procreation occurs only during winter months.

A right whale calf is born with callosities, but their pattern is not yet fully developed and they are not colonised by cyamids until the calf is several months old. Calves nurse for 6-12 months but often remain with their mothers for another two years. Females give birth to a new calf every three to five years.

NEW ZEALAND DISTRIBUTION

The southern right whale population in New Zealand's subantarctic appears to be steadily increasing at present and is currently estimated to be between 1300–2300 with around 50–60 calving females occupying Port Ross in the Auckland Islands each year. Around the mainland, however, recovery has occurred at less than snail's pace. The first 20th-century mainland sighting of a cow–calf pair occurred in 1991, and between 1991 and 2002 just 10 cow-calf pairs were recorded – it is not known if these represent 11 different mothers or if some sightings were of the same female that had returned a few years later with a new calf.

Part of the reason why there have been so few mainland sightings of females with offspring may lie in the reduced cultural memory within the whales themselves. Many whale species have a strong maternal fidelity to former calving areas and this instinct is passed on to their offspring. Young calves grow up following in the pathway of their mothers, from whom they learn about the migration routes that will lead them to traditional calving and feeding grounds. Because 19th-century whalers targeted the whale's traditional mainland calving grounds and so many cow–calf pairs were killed in mainland waters, the cultural memory of those calving areas was possibly wiped out along with the whale populations.

The 2011 discovery of nine DNA matches (seven females and two males), combined with the 2013 discovery of an additional three DNA and eight photo ID matches, between subantarctic and mainland southern right whales is an encouraging sign that the whales currently inhabiting waters to the south may be slowly moving northwards to colonise former right whale calving habitats.

White calves are usually males, and will grow up to become grey adults that retain a dark-spotted appearance.

An unanswered question still remains as to whether these subantarctic individuals are whales that are recolonising old haunts, or whether they are pioneers scouting for new calving grounds. In July 2012, a tiny southern right whale calf was sighted with its mother in Southland's Colac Bay. Although no one witnessed a birth, the calf's small size indicated that it was newly born, perhaps the first birth in mainland New Zealand waters since the end of the whaling era. On 21 August 2012, another southern right whale is believed to have given birth in Auckland's Browns Bay. The female was sighted coming close to shore in the morning, and by mid-afternoon she was observed with a calf. The two rare events are both encouraging and exciting signs that tohorā is slowly returning to the mainland.

THREATS AND CONSERVATION

The pre-whaling population of southern right whales is estimated to have been 60,000–120,000 worldwide. By 1920, as few as 60 mature breeding females may have survived the whaling onslaught. In New Zealand waters, the estimated pre-whaling population of 22,000–32,000 individuals was virtually wiped out by the 1850s, and by 1923 our total right whale population was estimated to number just 14–52 individuals. Recovery has been slow throughout the southern oceans but seems to be gaining momentum in recent years. Today's estimate of the total southern hemisphere population is around 12,000–15,000.

As more and more southern right whales return to mainland New Zealand's sheltered bays, they will face new threats. These include inshore fishing gear and other ocean debris, as well as ship strikes – the slow-moving animals have little capacity to get out of the way of boats.

In August 2012, researchers at the remote Auckland Island breeding grounds photographed a right whale with scars on its body that appeared to have been caused by a ship's propeller. The incident has led to a call for restricted ship speeds in Port Ross, the main calving and breeding grounds of the whale's New Zealand population.

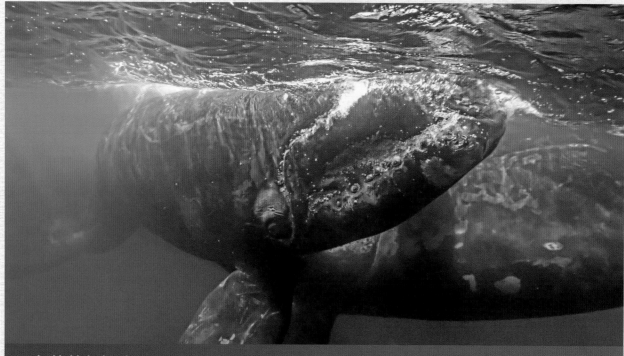

It is highly likely that this very young southern right whale calf was born in Aotearoa's mainland water, giving hope that the species is slowly returning to its former mating and calving grounds.

NEOBALAENIDAE (CETOTHERIIDAE): PYGMY RIGHT WHALE

Pygmy right whales are a reclusive species and until recently little has been known of their evolutionary history or exactly where they fit into the broader context. In November 2012, New Zealand scientists Ewan Fordyce and Felix Marx published revolutionary findings suggesting that the whales are, in fact, 'living fossils' belonging to the family Cetotheriidae, whose members were believed to have died out around 2 million years ago. Fordyce and other researchers extracted DNA from bone samples taken from whales that had stranded in New Zealand and molecular analysis of these samples and other material revealed features indicating that the pygmy right may be the last of the cetotheres species. Their results have led to the scientists formally referring the pygmy right whale to the family Cetotheriidae, in essence resurrecting that family from extinction and allowing science to study a whale species that exists today much as it and other family members may have done millions of years ago. If the scientists' recommendation is accepted, the family Neobalaenidae will no longer exist as the pygmy right whale is its only member.

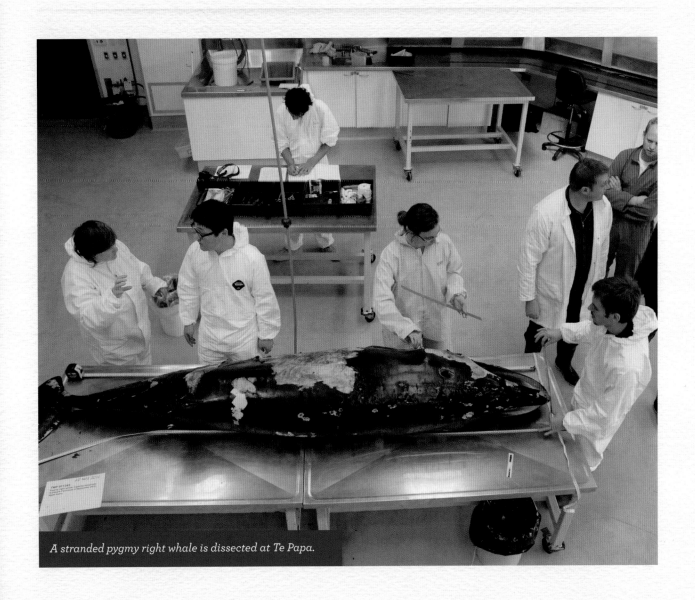

A stranded pygmy right whale is dissected at Te Papa.

Pygmy right whale

Caperea marginata (Gray, 1846)

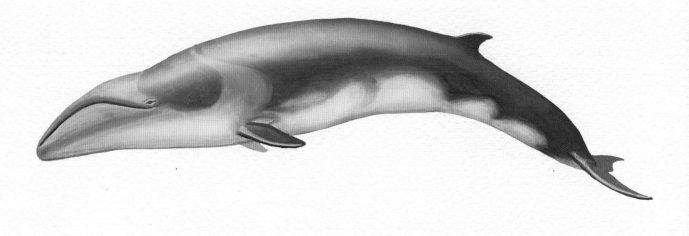

Size	Up to 6.5 m, up to 3400 kg
Gestation	Unknown
Birth size	1.6–2.2 m, weight unknown
Calving interval	Unknown
Lifespan	Unknown
Baleen	Up to 68 cm long, 210–230 per side; colour is yellowish white, often with a dark border; fringes are very fine and dense
Diet	Copepods and krill
IUCN status	Data Deficient
New Zealand population	Unknown

ETYMOLOGY

The pygmy right whale was first identified in 1846 from a piece of baleen, and since that time only a few dozen animals have been examined by scientists. Although 'right whale' is included as part of their name, they are not considered to be one of the right whale species and are consequently placed in their own separate family. Their genus name, *Caperea*, means 'wrinkle' in Latin and refers to the wrinkled appearance of the ear bone. The specific name, *marginata*, meaning 'enclosed with a border', refers to the dark border found around the baleen plates of most individuals.

DESCRIPTION

Pygmy right whales are the smallest and one of the least known of all the baleen whale species. They have an arched mouth like that of right whales, but the similarities to their much larger namesakes end there. The whales' dark grey body is small and sleek with narrow flippers, and unlike their larger cousins they possess a small, hooked dorsal fin.

One of the first comprehensive anatomical examinations of a pygmy right whale took place at Te Papa in 1996. Then in May 2008, leading whale scientists from New Zealand, Australia and the USA gathered at the museum for the dissection of a 2.3-metre juvenile that had been found stranded on a Northland beach the previous year. Local iwi Ngāti Kurī and Te Aupōuri gifted the whale's remains to Te Papa in order to give researchers the rare opportunity to study the anatomy of this unusual species, whose evolutionary history has only recently been revealed through DNA research.

From these dissections, scientists have learned that pygmy right whales are anatomical oddities. While most whales have an average of 10–15 pairs of ribs, pygmy rights have 18 ribs on each side, and when you get past the ninth rib, they flatten out and overlap, similar to the ribs of land-dwelling anteaters. Their larynx has an asymmetrical sac that differs from that of any other baleen whale, and raises the question of how they generate sound. All cetaceans possess a few hairs in specific areas, but the scientists performing the dissection found an unusually large number of hairs on the pygmy right whale's body.

BEHAVIOUR

Pygmy right whales are found only in the southern hemisphere and appear to have a circumpolar range. Most sightings or strandings are from Australia and New Zealand, with some animals also recorded from South Africa and South America. They inhabit temperate waters and have never been sighted south of the Antarctic Convergence (which is at latitude 60°S). Live sightings of these mysterious animals are uncommon; the whales have a low, indistinct blow and rarely display their dorsal fin or much of their body when they are on the surface. If their jawline is not visible at sea, they are often confused with minke whales.

NEW ZEALAND DISTRIBUTION

Between 1846–1999, there were only nine confirmed live sightings of pygmy right whales in Australasian waters. Fourteen pygmy right whales were sighted in 2001 about 450 kilometres southeast of New Zealand, but there have been no confirmed sightings of live animals around the mainland that have not resulted in strandings. The majority of pygmy right whale strandings in New Zealand have occurred at Stewart Island/Rakiura, Golden Bay, in the Cook Strait, around Auckland and the Bay of Plenty. At least seven pygmy right whales have stranded in Golden Bay since 1999. The most recent occurred in 2012 when three separate whales stranded. One died and two were returned to sea.

THREATS AND CONSERVATION

There is little information on the abundance of pygmy right whales, which are consequently classed as Data Deficient by the IUCN. They were never targeted by whalers and there is no reason to believe their populations are endangered aside from the current environmental threats facing all cetacean species.

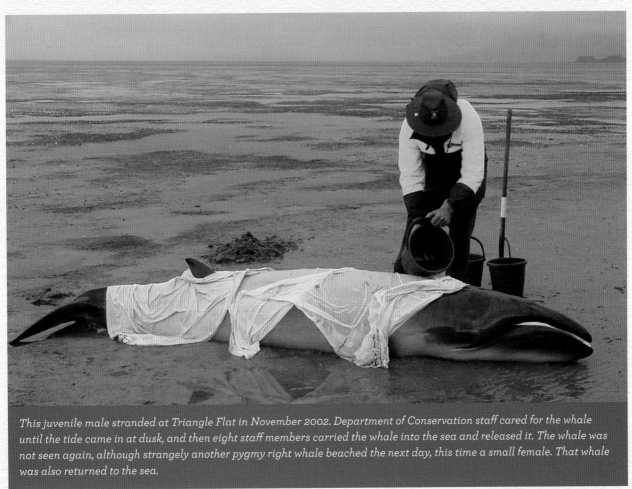

This juvenile male stranded at Triangle Flat in November 2002. Department of Conservation staff cared for the whale until the tide came in at dusk, and then eight staff members carried the whale into the sea and released it. The whale was not seen again, although strangely another pygmy right whale beached the next day, this time a small female. That whale was also returned to the sea.

BALAENOPTERIDAE: RORQUALS

All balaenopterids – commonly known as the rorquals – have ventral throat pleats that expand when the whales are feeding. Seven of the eight rorqual species and the two subspecies known in New Zealand waters share the genus name *Balaenoptera*, which means 'winged whale', while the eighth (the humpback) has a separate genus name, *Megaptera*, which means 'large wing'. Balaenopterids possess a dorsal fin and all have short heads that are less than a quarter of their body length. Their upper jaw is unarched and, compared to balaenids, their baleen is short and wide. Most species use a gulping technique when they are feeding, although the sei whale is also known to skim for its food. Only one Balaenopteridae species, Bryde's whale, is resident in New Zealand; the rest are migrants that mainly travel through our waters in spring on their way to summer feeding grounds in higher latitudes, and in autumn on their way to winter breeding grounds in lower latitudes. While these may be the primary seasons when migrant rorquals occur in New Zealand waters, the whales may also occasionally be observed at other times of the year.

A humpback calf-cow pair (top) with an escort.

Antarctic blue whale

Balaenoptera musculus intermedia (Burmeister, 1871)

Size	Up to 33.6 m, up to 190,000 kg
Gestation	10–12 months
Birth size	7–7.9 m, 2700–3600 kg
Calving interval	2–3 years
Lifespan	At least 80–90 years
Baleen	Up to 1 m long and 50–55 cm wide, 260–400 per side; normally black in colour but can vary and sometimes turn grey in older whales
Ventral pleats	60–88, extending to the navel
Diet	Primarily krill
IUCN status	Critically Endangered
New Zealand population	Migrant

ETYMOLOGY

The blue whale was first described in 1694 by physician and naturalist Robert Sibbald from a male that stranded in Scotland's Firth of Forth. Carolus Linnaeus later assigned the species the scientific name *Balaenoptera musculus*, the specific Latin name meaning 'muscle'. The whale's mottled bluish-grey colour led to its English common name blue whale, from the Norwegian *blåhval*. One of the species' better-known nicknames comes from author Herman Melville, who noted the yellowish tinge on the whale's underbelly (caused by micro-organisms such as the diatom *Cocconeis ceticola* and other marine algae) and called it the sulphur-bottom whale. The Antarctic blue whale, *B. musculus intermedia,* was recognised as a southern hemisphere subspecies by German zoologist Hermann Burmeister in 1871.

DESCRIPTION

Blue whales are the largest creatures ever to have lived. The two longest known individuals, measured by whalers, were 33.6 metres and 33.3 metres long. A blue whale's flippers are 3–4 metres long, while its tail flukes are up to 8 metres wide; in comparison, its dorsal fin is tiny and only briefly visible when the whale dives. Weight can be difficult to determine, early scientists would cut a whale into chunks and weigh each piece. One of the heaviest recorded blue whales, measured by American scientists, weighed over 180 tonnes.

BEHAVIOUR

Blue whales are primarily lunge-feeders and appear to dine almost exclusively on euphausiids (krill). Although some feeding is done at or near the surface, the whales will also dive into large concentrations of their prey that are as deep as 100 metres, gulping in huge mouthfuls of water and krill and then rising to the surface to expel the water. Feeding is often observed in early morning or late evening when prey is closer to the surface.

Resting whales will exhale, or 'blow', every 10–20 seconds and will sometimes lift their tail flukes prior to a dive, but at other times they will just slip under the water. They can descend to depths greater than 150 metres and remain underwater for 20 minutes or longer, although average dive times are eight to 15 minutes. When feeding, the whales' average speed is 3–6 kilometres per hour, but they have been known to reach speeds of 30–35 kilometres per hour when being chased. Juveniles have been observed breaching but adults rarely do so – propelling all that weight clear of the water can't be easy!

Blue whales are sexually mature at around 5 to 15 years of age. The mating season extends from late autumn into winter, and females appear to give birth every two to three years after a 10–12-month gestation period.

When they are travelling, the whales are normally seen alone or in pairs, but on their feeding grounds loose aggregations sometimes gather in areas where there are high concentrations of krill. Although they are known to migrate into warm waters in winter, their exact mating and calving grounds are virtually unknown.

NEW ZEALAND DISTRIBUTION

In New Zealand, Antarctic blue whales are normally sighted as they pass by on their journey south to summer feeding grounds in cold polar waters or as they travel north to tropical waters to breed. Recent research off the lower west coast of the North island, however, suggests blue whales may be using the South Taranaki Bight as a feeding ground. There have been numerous sightings of the whales feeding on zooplankton in the bight. If this is confirmed, their status as a migrant species in New Zealand will need to be re-evaluated and future means of protecting the species from ship strikes and oil spills will need to be assessed.

RECORD-BREAKER

The blue whale's size isn't its only record-breaking feature; check out these fascinating facts:

- The blue whale's sounds can be heard up to 1000 kilometres away and range in frequency between 10 hertz and 40 hertz, making them the lowest on the planet.

- The blue whale's heart is the size of a small automobile and can weigh up to 640 kilograms. The aorta artery is about 23 centimetres in diameter, large enough for a small child to crawl through.

- The whale's 2.7-tonne tongue weighs more than a small elephant, and when its ventral pleats are expanded the mouth can hold up to 90 tonnes of food and water. Blue whales consume up to 3600 kilograms of krill in a single day, equivalent of around 40 million of these tiny crustaceans.

- Blue whales have the world's largest babies: newborns are 7–8 metres long and weigh up to 2700 kilograms, the same weight as an adult hippopotamus. During its first seven months of life, the calf drinks up to 400 litres of milk and gains 70–90 kilograms every day, or more than 3 kilograms per hour.

- A blue whale's lungs can hold up to 5000 litres of air and, when it exhales, its spout reaches 9–12 metres high – about the height of a power pole.

A blue whale skull is installed in the Smithsonian Institute.

THREATS AND CONSERVATION

The powerful blue whales are fast swimmers, reaching up to 35 kilometres per hour in short bursts, which saved them during the early whaling years, when sailing ships were used. Once the harpoon gun and steamships were introduced, however, their numbers soon plummeted. The first Antarctic blue whales were taken off South Georgia in 1904–5. By 1925, the catch had increased dramatically and, in 1930–31 alone, more than 29,000 were taken off Antarctica. By the time blue whale hunting was banned in 1966 and illegal Russian whaling ended in the early 1970s, at least 330,000 blue whales had been killed in Antarctic waters, along with another 33,000 in the rest of the southern oceans. The current estimated population of 5000 Antarctic blue whales is still believed to be less than 3 per cent of its original numbers. Scientists have theorised that the known incidents of blue/fin whale hybridisation (11 have been documented) are an indication that there is a lack of breeding partners of their own kind.

While whaling for blue whales has ceased, they continue to face the same threats encountered by other cetacean species: pollution; vessel strikes; increased ocean noise, which makes it difficult for them to communicate over vast distances; and global warming, which threatens krill, their primary food source, in cold polar waters (krill stocks may have been reduced by up to 80 per cent since 1980, possibly due to ice melt through global warming, reducing the availability of the ice algae on which the krill feed).

Blue whale calves nurse for around six to eight months; by the time they are weaned, they have almost doubled in size to reach a length of 14–15 metres.

When surfacing to breathe, the blue whale raises its blowhole out of the water to a greater extent than fin or sei whales. The whale is so long that its head normally disappears under the surface before its dorsal fin appears. These traits help observers to differentiate between species out at sea.

Pygmy blue whale

Subsp. *Balaenoptera musculus brevicauda* (Ichihara, 1966)

Size	Up to 24 m, weight unknown
Gestation	About 11 months
Birth size	Unknown
Calving interval	Unknown
Lifespan	Unknown
Baleen	About 50 cm long, about 320 per side; mostly black
Ventral pleats	Unknown
Diet	Mainly krill
IUCN status	Data Deficient
New Zealand population	Migrant

ETYMOLOGY

The pygmy blue whale, a subspecies of the blue whale, was first described in 1966 by Japanese scientist Tadayoshi Ichihara, who gave it the subspecies epithet *brevicauda*, meaning 'short-tailed' in Latin.

DESCRIPTION

The tadpole-shaped pygmy blue is smaller and has a shortened tailstock compared to its larger and more streamlined Antarctic blue whale relative (see page 257). Its baleen is shorter and broader, and the shape of the blowhole is different. Despite the difference in size, the two blue whales can be difficult to distinguish at sea as their bodies both display small dorsal fins and mottled grey-blue skin, although pygmy blues are often darker and more wrinkled.

BEHAVIOUR

Little is known about the breeding and social behaviour of pygmy blue whales. Distinctions between this subspecies and the larger Antarctic blue whales were not often made on whaling vessels, so no specific data on the animals have been recorded. Like their larger cousins, they tend to be solitary except on feeding grounds, where numerous animals may gather near large concentrations of food. They are known to be lunge-feeders whose main prey is krill. It is believed their biology is similar to that of the larger subspecies of blue whale, but further studies are needed to determine just how similar the two species are.

The calls or songs of pygmy blues appear to be more complex than the call types of other blue whale subspecies. Recordings in the Indian Ocean suggest there may be a minimum of three distinct acoustic populations of pygmy blues living

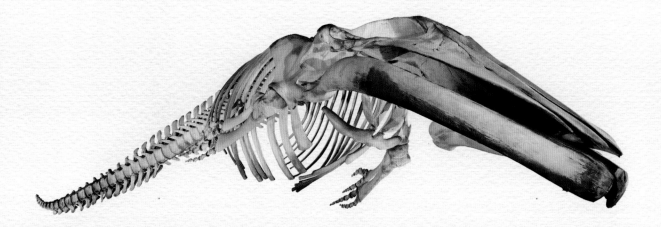

The skeleton of a 20.6-metre sub-adult male pygmy blue hanging in Te Papa is from a whale that was struck by a container ship off the North Cape in 1994. Numerous fresh marks on the whale's body suggested that it had been attacked by orca and may have been resting at the surface when it was hit by the ship.

in the region. The sounds of pygmy blues and Antarctic blues not only differ from one another, they also differ from the sounds made by northern hemisphere blue whales. Evidence suggests that only males produce the songs, but this is yet to be proven.

NEW ZEALAND DISTRIBUTION

The migratory journeys of pygmy blues vary, with some whales travelling to Antarctic waters, some to the subantarctic and some, such as those in the northern Indian Ocean, staying put as resident populations. While live sightings of pygmy blue whales in New Zealand waters can be difficult to confirm, stranded animals have provided evidence of their presence here.

It is very difficult to distinguish between Antarctic and pygmy blue whales at sea. Recent sightings of Antarctic blue whales in the Taranaki Bight off the North Island's west coast may include, or may be, pygmy blue whales.

THREATS AND CONSERVATION

The exact number of pygmy blues that were killed by whalers is unclear, but it is believed that many of the illegal Russian catches of blue whales were, in fact, of pygmy blues. Pollution and ocean debris are known to be a particular problem for these whales. In 2009, a pygmy blue whale that stranded on the South Island's west coast was found to have rope lodged in its throat, while a 22.5-metre pygmy blue that stranded in May 2011 along the North Island's west coast was extremely emaciated when it came ashore and may also have swallowed ocean debris.

Fin whale

Balaenoptera physalus (Linnaeus, 1758)

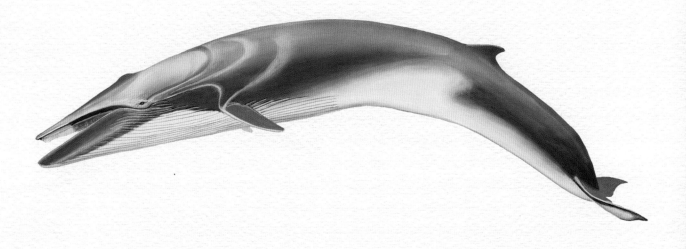

Size	Up to 27 m, up to 80,000 kg
Gestation	11–12 months
Birth size	6–6.5 m, 1800–2700 kg
Calving interval	2–3 years
Lifespan	0–90 years
Baleen	Up to 70 cm long, 260–480 per side; colour is dark grey to black; fringes appear yellowish white to olive green
Ventral pleats	50–100 pleats, extending from throat to navel
Diet	Krill, schooling fish and, occasionally, small squid
IUCN status	Endangered; Mediterranean sub-population Vulnerable
New Zealand population	Migrant

ETYMOLOGY

The species name *physalus* roughly translates as 'bellows' and is thought to refer either to the whale's ventral pleats, which puff up or bellow out during feeding, or to its highly visible blow. The whale's common name is a reference to its prominent dorsal fin.

DESCRIPTION

Fin whales are the second largest whale species, and southern hemisphere individuals are 3–4 metres longer than their northern hemisphere counterparts. One of their most distinctive features is the asymmetrical colouration on their heads: on the right side the lower jaw, mouth cavity and some of the baleen is white; on the left side the colour is uniformly grey or black.

BEHAVIOUR

Fins are found worldwide in tropical, temperate and polar waters. These so-called greyhounds of the sea are sleek and streamlined, capable of bursts of speed in excess of 37 kilometres per hour. One tagged individual made a marathon swim and was monitored travelling over 3700 kilometres at an average speed of 17 kilometres per hour. Tagged whales have been recorded diving to depths of 550 metres.

Although little is known about the species' reproductive behaviour, it is believed sexual maturity is reached between six and eight years and that most mating takes place during winter. Calves are born after an 11–12 month gestation and are nursed until they are around six or seven months old. Calving appears to occur every two to three years. Fin whales are known to interbreed with blue whales and several hybrid calves have been identified, with at least one of those calves becoming pregnant herself.

Fin whales make a variety of low-frequency (20-hertz) vocalisations, which are emitted in consistent patterns during the breeding season and can be heard hundreds of kilometres away. The animals are usually seen alone or in small groups of three to seven individuals, and are known to migrate into higher latitudes for feeding and into lower latitudes for mating, although no specific breeding or calving grounds have been identified to date.

NEW ZEALAND DISTRIBUTION

In New Zealand, confirmed sightings of fin whales are relatively rare owing to their offshore migratory routes and also to the difficulty of differentiating the species from other large cetaceans such as blue and sei whales. The majority of confirmed sightings have come from Northland and from offshore waters around the Bay of Plenty.

THREATS AND CONSERVATION

Prior to the invention of the steam engine and the explosive harpoon, the speedy fin whales were too fast for whalers. Things changed in the 20th century, however, when fins gained the unhappy distinction of being the most hunted species – over 720,000 were killed in the southern hemisphere alone between the late 1800s and the early 1970s. Although the species was protected in 1975, it is still considered to be endangered, particularly in the southern hemisphere. Chemical pollution, interaction with fisheries, boat strikes and increased human disturbance

such as underwater acoustic testing are among their main threats today. Worldwide, fin whales are taken in Greenland (around 19 per year), Iceland (which sets its own quota of 154 per year) and in the Antarctic, where Japanese whalers resumed catching the species in 2005 when they established an 'experimental' capture programme that takes 3 to 10 individuals a year.

On average, fin whales blow two to five times at intervals of 10–15 seconds while on the surface and will then dive for 5–15 minutes, although they are capable of remaining underwater for longer periods of time.

In 1882, a large whale skeleton was found on a Nelson beach by Captain Jackson Barry. Captain Barry toured the South Island with the skeleton, charging people a sixpence to view the bones of the 'magnificent creature'. In 1884, he arrived in Dunedin and sold the skeleton to the Otago Museum. Although a few small bones had 'gone missing', Dr Bourne of the University of Otago was able to replace them with wood replicas. The skeleton, measuring 16.7 metres, is believed to be from a juvenile fin whale and is one of the few virtually complete skeletons of the species in the world.

Sei whale

Balaenoptera borealis (Lesson, 1828)

Size	14–18 m, up to 40,000 kg
Gestation	11–12 months
Birth size	4.5–4.8 m, about 680 kg
Calving interval	2–3 years
Lifespan	At least 60 years
Baleen	Less than 80 cm long, 300–400 per side; black in colour; very fine fringes of light grey to white
Ventral pleats	32–65 short pleats, ending well before navel
Diet	Copepods, krill, small fish and squid
IUCN status	Endangered
New Zealand population	Migrant

ETYMOLOGY

The sei whale is split into two subspecies; the southern hemisphere subspecies is *Balaenoptera borealis schlegelii* – the species name, *borealis*, means 'northern', and the subspecies name, *schlegelii*, was given in honour of the German zoologist H. Schlegel. The whale's common name, sei (pronounced 'sigh'), is the Norwegian word for 'pollock'. Both the fish and the whales arrive off Norway's coasts at the same time each year in order to feed on abundant shoals of plankton, and as a result the whales became known as sei whales. The whales have also been commonly known as lesser fins, as both they and fin whales display prominent dorsal fins and are difficult to distinguish at sea.

DESCRIPTION

Sei whales are also similar in appearance to Bryde's whales (see page 266) and the two species have been confused for many years. The main distinguishing characteristic is the longitudinal ridges on the heads of both whales: sei whales have only one ridge, while Bryde's have three. The southern sei whale is larger than the northern subspecies and while most average around 14–16 metres in length, one overly large female measured 20 metres and weighed almost 45 tonnes.

BEHAVIOUR

Sei whales, the 'cheetahs' of the sea, are fast swimmers but are capable of only short bursts of speed in excess of 45 kilometres per hour. They have one of the most varied diets of any baleen whale and alter their feeding techniques to match their prey. When small fish and krill are on the menu, they engage in lunging or gulping techniques, while at other times they use a skimming method, similar to that of right whales (see page 249), and feed on copepods and other tiny prey.

Seis appear to avoid extreme polar and tropical environments, and migrate instead between sub-polar and subtropical waters. While they are found throughout the world's oceans, their exact distribution is poorly known as they live far from land and change areas on an irregular basis.

NEW ZEALAND DISTRIBUTION

In New Zealand, sei whales are sighted only on occasion, with most reports coming from the east coast of the North Island, particularly around the Bay of Plenty. In November 2012, a baby sei whale beached in the Motupipi estuary in Golden Bay. The whale was so young that it had to be euthanased as it would not have survived at sea without its mother.

THREATS AND CONSERVATION

The streamlined sei whales were not initially targeted by whalers, but this changed in the 1960s as blue, fin and humpback whale populations declined. In 1964–65 alone, more than 25,000 seis were taken globally, and by 1978 the species had been 'fished' out, with a paltry 150 captured between 1978 and 1979. In the southern hemisphere, more than 152,000 sei whales were killed between 1910 and 1979. Small numbers of sei whales continue to be hunted today by both the Japanese (in the North Pacific) and Iceland (in the North Atlantic) under the International Whaling Commission's scientific research programme. In 2010, a restaurant in California was found to be serving sei whale meat to its customers.

This juvenile sei whale stranded in November 2012, presumably after losing its mother.

Other names for sei whales include pollack whale, coalfish whale, sardine whale, Japan finner and Rudolph's roqual!

Bryde's whale

Balaenoptera edeni (Anderson, 1869)

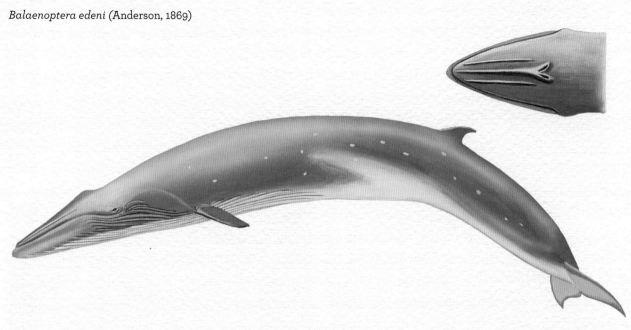

Size	Up to 16.3 m, up to 35,000 kg
Gestation	About 12 months
Birth size	4–4.5 m, 1000 kg
Calving interval	2 years
Lifespan	Unknown
Baleen	Up to 50 cm long and 19 cm wide, 250–370 per side; usually black or dark grey, although sometimes creamy white, especially near rostrum
Ventral pleats	40–70, reaching near or beyond navel
Diet	Mainly schooling fish; occasionally krill and squid
IUCN status	Data Deficient
New Zealand population	Resident year-round; Nationally Critical

ETYMOLOGY

Bryde's (pronounced 'bru-das') whales have a complicated and as yet unresolved taxonomy.

The whale was first described from a stranded specimen found off the coast of Burma in 1878 by Scottish zoologist John Anderson, who named it *Balaenoptera edeni* in honour of Ashley Eden, the British high commissioner to Burma. Then, in 1913, a larger form of the whale was discovered off the coast of South Africa and was given the scientific name *B. brydei* and common name Bryde's whale by Norwegian scientist Ørjan Olsen to honour Johan Bryde, a pioneer of the South African whaling industry. In the 1950s, the two whales were grouped together as *B. edeni*, but retained the common name Bryde's. Genetic research in 2006, however, indicated that the two are indeed distinct but closely related species, although the two forms are yet to be officially separated and

given full species status. A pygmy form, Omura's whale, found in the eastern Indian and western Pacific oceans, was accepted as a separate species, *B. omurai*, in 2003. Genetic analysis on New Zealand's resident population suggests the whales belong to the *B. brydei* group, but until they are formally given species status, they remain classed as *B. edeni*.

DESCRIPTION

For many years, Bryde's whales were confused with sei whales (see page 264). They are distinguished from sei and other rorquals by the three longitudinal ridges running from the tip of the snout to the blowholes, although the ridges may be difficult to observe at sea (detail above). The whales have a dark back with light grey or creamy-white undersides that sometimes have a pinkish tinge. Although the whales usually arch their back prior to a dive, they seldom raise their tail flukes above the water.

The three longitudinal ridges on the head of the Bryde's whale distinguishes it from other rorqual species such as sei, minke and fin with which it can be confused at sea.

The waters of the Hauraki Gulf provide many dining opportunities to a number of marine mammal and seabird species including the Bryde's whale.

Like all rorquals, Bryde's have throat pleats that expand when the whales are feeding.

BEHAVIOUR

Most Bryde's whales have a fairly restricted home range; they prefer tropical and temperate waters and do not migrate into cold polar seas. Preliminary aerial surveys carried out in the Hauraki Gulf and northeastern New Zealand waters between 1999 and 2003 by Alan Baker and Bénédicte Madon led to on-going research from the air and at sea by many other scientists, which has provided new information on the species' distribution and behaviour. The research has revealed that Bryde's are using the Hauraki Gulf as both a feeding and a nursery ground. New calves are observed year-round, with peak births occurring in late summer and early autumn. Calves are 4–4.5 metres at birth and nurse for around six months; by the time they are weaned, they are around 7 metres long.

Whales within the gulf are normally seen singularly or in pairs, although loose aggregations of five or more may be observed around feeding areas. Both aerial and vessel surveys used GPS technology to pinpoint the whales' locations and to track their movements, and the boat researchers also used photo identification and DNA analysis. No fewer than 72 individual whales have been identified in the gulf from dorsal fin markings, and researchers estimate that the local population numbers from 159 to 200 individuals.

Bryde's feed year-round in the Hauraki Gulf. Foraging occupies around 50–70 per cent of the whales' time, and while they sometimes feed on krill, their main diet consists of schooling fish such as anchovies and pilchards. The highly productive waters attract a large diversity of marine life and Bryde's are often observed lunge-feeding amidst shoals of small fish alongside dolphins, sharks, diving gannets and other seabirds. Bryde's do not spend a lot of time lying on the surface; rather, they travel slowly when they are not foraging. Aside from feeding aggregations and mother–calf pairs, the whales appear to be rather solitary.

NEW ZEALAND DISTRIBUTION

Bryde's whales were first recognised in New Zealand waters in the late 1950s when samples taken from what were thought to be sei whales were examined by William Dawbin at the Great Barrier Island (Aotea Island) whaling station. Today, they are the most frequently sighted large whale in northern New Zealand. The 1999–2003 aerial study led by Alan Baker examined the distribution of the Bryde's whale population in New Zealand waters. The study confirmed the species' presence along the northeast coastal regions between East Cape and North Cape, and the fact that Bryde's live year-round throughout the Hauraki Gulf. The majority of Bryde's strandings (at least 25) have occurred between the Bay of Islands in the north and the Coromandel Peninsula (Firth of Thames) in the south.

THREATS AND CONSERVATION

The majority of New Zealand's Bryde's whales reside at Auckland's back door in the inner Hauraki Gulf, where they compete for space with the 2000-plus large ships that enter and leave the Port of Auckland every year.

In addition, they face the threat of boat strike from fast ferries and as many as 150,000 recreational boats. Between 1996 and 2013, there were 43 reported Bryde's whale deaths in the gulf, of which 16 were known to have been caused by vessel strikes. Shipping representatives working in conjunction with the Department of Conservation and whale scientists have agreed to contribute funding for research into ways of avoiding ship strikes and to implement a number of measures to help safeguard the whales which include:

- reducing vessel speed

- establishing shipping lanes outside the areas in which whales are sighted

- assigning crew to watch for whales during daylight hours while ships move through the gulf

- establishing a Hauraki Gulf large whale warning system so that reports of large whales within the gulf are relayed to ships.

The small population size of resident Bryde's whales in New Zealand has resulted in them being classified by the Department of Conservation as 'Nationally Critical'.

Aquaculture facilities can be killers – in this 1996 stranding on Great Barrier Island (Aotea Island), a Bryde's whale was found dead after mussel spat line was entangled in its mouth.

Antarctic minke whale

Balaenoptera bonaerensis (Burmeister, 1867)

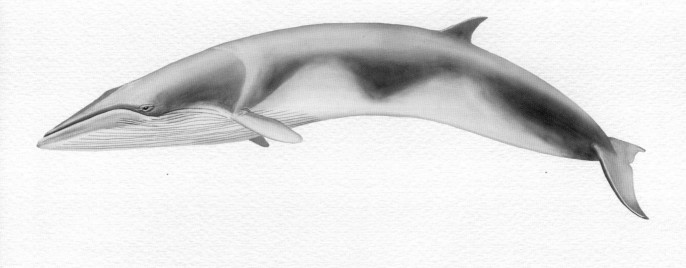

Size	8.5–10.7 m, about 9000 kg
Gestation	10–11 months
Birth size	About 2.8 m, about 450 kg
Calving interval	1–2 years
Lifespan	Unknown
Baleen	20–30 cm long, 200–300 per side; asymmetrical colouration – most are dark grey, but many anterior plates (especially on the right side) are white
Ventral pleats	22–38, extending just past flippers
Diet	Mainly krill; some schooling fish
IUCN status	Data Deficient
New Zealand population	Migrant

ETYMOLOGY

Antarctic minke whales were officially recognised as a separate species in the late 1990s, following DNA analysis. Their specific name, *bonaerensis*, means 'Buenos Aires', which is where the first described specimen was found. The minke whale's common name comes from Norway where the story has it that a whale spotter named Meincke mistakenly identified a minke as a blue whale, and after that the much smaller roquals were sarcastically called 'Minikies' whales.

DESCRIPTION

Genetically, the larger Antarctic minke have been found to be more closely related to sei and Bryde's whales (see pages 264 and 266, respectively) than they are to the common or dwarf minke (see page 271). The whale's flippers are a uniform light grey colour and sometimes have a dark band around them, but they lack the distinctive white patch found on dwarf minke flippers.

BEHAVIOUR

Knowledge about the Antarctic minke's biology and behaviour is minimal. Seasonal sex and age segregation occurs in minke whales in the northern hemisphere, but research on the social behaviour of southern hemisphere minkes is limited. The whales are believed to be sexually mature at around six to seven years. Calves are born after a 10–11 month gestation and are weaned at around five or six months. Females appear to be pregnant around 90 per cent of the time, indicating an annual reproductive cycle. The winter breeding areas of the species remains a mystery, with no specific mating or calving grounds documented to date.

The Antarctic minke has a circumpolar range, spending summer months feeding in Antarctic waters. During winter, it appears that some

Minke whales have been observed near the icepack in late summer and early autumn and some animals may even overwinter in Antarctic waters.

populations migrate to warmer waters between 7°S and 35°S, while other populations remain around the Antarctic Convergence and may even overwinter in Antarctic waters.

NEW ZEALAND DISTRIBUTION

In New Zealand waters, Antarctic minke are known mainly from stranded animals. One of the latest strandings occurred in mid-August 2012 in Northland, when a young male came ashore at Patea Beach. The whale was refloated but later returned to shallow waters and ultimately had to be euthanased.

THREATS AND CONSERVATION

It is estimated that more than 100,000 Antarctic minkes have been killed in the past century. Although the whales were not initially the focus of large-scale commercial whaling, the species was targeted when other whale stocks became depleted, and from 1979 to 1986 they were the predominant species killed by Southern Ocean whalers. Japan continues to kill several hundred Antarctic minkes a year as part of its 'scientific research' programme. There is considerable uncertainty about the abundance and trends in Antarctic minke whale populations but there is broad consensus that the population lies between 300,000 and 500,000. Antarctic minke whales are also heavily targeted by orca in Antarctic waters.

Dwarf minke whale

Subsp. Balaenoptera acutorostrata

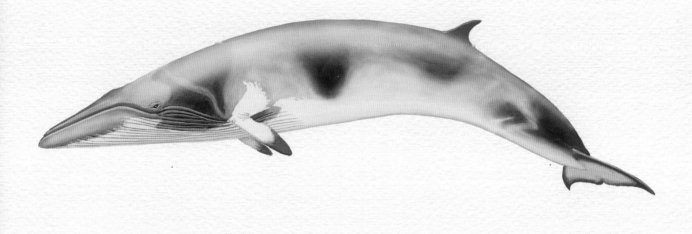

Size	7–8 m, 6400 kg
Gestation	About 10–11 months
Birth size	About 2 m, weight unknown
Calving interval	Probably 1–2 years
Lifespan	Unknown
Baleen	Up to 25 cm long, 200–300 per side; light coloured
Ventral grooves	About 50–60, extending just past flippers
Diet	Krill and schooling fish
IUCN status	Least Concern
New Zealand population	Migrant

ETYMOLOGY

In 1981, Australian diver Rob Prettejohn did a drawing of an 'unusual form' of minke that he encountered during a dive on the Great Barrier Reef. The whale he encountered was given the name 'dwarf minke' in 1985 by whale biologist Peter Best, who was examining carcases of the same form of minke at a whaling station in South Africa.

The diminutive dwarf minke has since been recognised as a form of the common minke (found in the northern hemisphere) and may soon be accorded separate subspecies status. At the time of writing, however, the dwarf minke was still listed as an unnamed subspecies. Dwarf minkes are the second-smallest baleen whale (after the pygmy right whale) and are only found in the southern hemisphere. The common minke whale's sharply pointed snout led to its specific name *acutorostrata*, meaning 'sharp-snouted', as well

as one of its often-used nicknames, piked whale.

DESCRIPTION

Dwarf minkes resemble both common and Antarctic minkes (see page 269), but their colour pattern differs. Their most distinctive feature is their flippers, which display a white patch that is tinged with dark grey at the tip, similar to the flippers of the northern hemisphere form (the patch is absent in the Antarctic minke). Dwarf minkes also have a dark patch on the side of the throat that extends from the eye to the flipper. In 1983, New Zealand cetacean expert Alan Baker was among the first scientists to identify the dwarf form in New Zealand waters from the examination of minke whales that had stranded here.

BEHAVIOUR

Dwarf minkes appear to live alone or in small groups, but aside from that little is known about their social structures or mating grounds. The whales are believed to have a circumpolar distribution in the southern hemisphere.

Between March and October, the whales are commonly sighted off Australia's Great Barrier Reef. In the 1980s, dwarf minkes began to approach dive boats and divers on the reef. The encounters have since increased and intensified, with the inquisitive whales now approaching boats and swimmers on a regular basis – sometimes as many as a dozen at a time interact with swimmers. Interactions are initiated by the whales and can last for hours; in fact, it is usually the boats and divers that break off the encounter.

NEW ZEALAND DISTRIBUTION

In summer, sightings of what are believed to be dwarf minkes occur in subantarctic waters south of New Zealand and Australia, and they have also been observed close

to the ice edge in Antarctica. The majority of reported mainland New Zealand sightings come from the Hauraki Gulf, Bay of Plenty and northeast coast of the North Island, and from Dunedin and Southland in the South Island.

THREATS AND CONSERVATION

Although dwarf minkes were hunted by commercial whalers, confusion with the Antarctic minke means there are few reliable records about the actual numbers taken. Some dwarf minkes are known to be taken along with the Antarctic minke catch by Japanese whalers under the International Whaling Commission's scientific whaling permit. The 2008 IUCN Red List of threatened species reported that of 1700 minke whales taken by Antarctic pelagic whaling between the 1987–88 and 1992–93 seasons, 16 are known to have been dwarf minke. One was taken at 65°S and the remainder between 55°S and 62°S, the northern limit of the whalers' operations.

A stranded dwarf minke whale.

Humpback whale, Paikea

Megaptera novaeangliae (Borowski, 1781)

Size	Up to 17 m, up to 40,000 kg
Gestation	12 months
Birth size	About 4.3 m, about 680 kg
Calving interval	2–3 years
Lifespan	At least 50 years
Baleen	50–60 cm long and 34 cm wide, 270–400 per side; usually dark grey to black but sometimes pale grey or white near jaw tip
Ventral pleats	14–35, extending to navel and beyond
Diet	Krill and schooling fish
IUCN status	Least Concern; Arabian Sea and Oceania sub-populations Endangered
New Zealand population	Migrant

ETYMOLOGY

Humpback whales have ventral throat pleats and share many characteristics with the other rorqual species, but they also possess distinctive features that led to them being placed in their own separate genus, *Megaptera*, which means 'big wing' and refers to their exceptionally long flippers. Their specific name, *novaeangliae*, means 'New England', which is where the whales were first described and named. To Māori, humpback whales are known as paikea, while the species' common English name comes from the whale's habit of rounding its back, or humping up, prior to diving.

DESCRIPTION

The humpback's head is covered with tubercles, bumps or knobs, each with a stiff hair growing out of it somewhat like a dog's whisker. The function of these whiskered knobs is unknown although it is thought to be sensory, perhaps helping the whales detect the presence of schooling prey. In addition to the tubercles, humpbacks' bodies are often encrusted with barnacles, particularly around the edges of the tail flukes and flippers. The humpback's remarkable flippers average 5 metres long and are almost a third the length of the whale's entire body. They are longer than those of any other whale species, and in fact are the longest known appendages of any animal species.

BEHAVIOUR

Of all the large whales, humpbacks are the most active, putting on spectacular displays of breaching, lob-tailing, spyhopping and flipper-slapping. At times their bodies completely clear the water when they breach; at others, they do multiple breaches in a row – pretty impressive when you stop to think that they are bringing the weight of a large concrete mixer out of the water with every leap.

While all of this action may look like it is just for fun, there are undoubtedly

Humpback flippers are longer than those of any other whale species and in fact are the longest known appendages of any animal species.

reasons behind the behaviours. A spyhopping whale is likely to be checking out its surroundings above the surface, flipper-slapping may indicate a whale's location, while a noisy breach or tail-lob may be a way for a whale to communicate its presence or intimidate other whales by sending out a signal of strength. On their feeding grounds, humpbacks have been observed using their flippers or tails to whack schools of fish, while on their mating grounds, males sometimes whack competitors with their tails.

Humpbacks undertake some of the longest migrations of any whale species. Their annual journey in southern oceans goes from their cold Antarctic feeding grounds to their mating and calving grounds in tropical waters at latitudes around 20°S and then back again – a round trip of up to 15,000 kilometres. The whales do not feed on their mating and calving grounds, so they must put on as much weight as possible when they are in polar waters. Because their main prey in the southern oceans is krill, southern hemisphere humpbacks seldom engage in the elaborate bubble-net strategy observed in their northern hemisphere counterparts (see page 48) but instead use lunge-feeding techniques. While the whales may work as a cooperative group during feeding, their migration to warmer waters is usually undertaken alone or in pairs. Pregnant females are usually the first to arrive on the feeding grounds; they are also the last to leave, as they must build up the fat reserves they will need when they give birth in the tropics.

Once the whales arrive on their mating grounds, the males sing – although it is not clear whether this is to attract females, to establish their location or to impress other males. When they are not singing, the males often chase other males, pushing, shoving, bellowing and blowing as they do so; even headbutting another male is not unheard of. A single female may be followed by an aggregation of up to 20 males, all vying for her attention.

While the mating activities are taking place, pregnant females are searching for sheltered areas where they can give birth and care for their young. New mothers and their young will be among the last to leave the calving grounds. By the time she leaves the tropics, a nursing mother will have lost up to a third of her body weight. The new calf will only be a few months old when it sets off on its first big adventure. It will make only a few journeys by its mother's side and then will be left to swim the long distances on its own. During its lifetime, the combined distance a humpback covers is equivalent to travelling to the moon and back!

NEW ZEALAND DISTRIBUTION

The humpback whale's migrations are not only long, they are also predictable. This predictability made it easy for whalers to target the areas where the whales would be at certain times of the year. New Zealand's humpbacks migrate up the east coast and through Cook Strait in winter on their way to their tropical mating and calving grounds, and in summer they head south again past the country's west coast to their Antarctic feeding grounds. Humpbacks in the South Pacific are once thought to have numbered up to 50,000–60,000, with 20,000–30,000 migrating along the eastern Australian coast and about 20,000 or more breeding in the remainder of Oceania. These populations were particularly hard hit from 1959 to 1961, when over 25,000 whales were killed, and from 1961 to 1962, when another 13,000 were killed by illegal Soviet whaling (see page 130). What is unknown is how many humpbacks were illegally taken by other whaling nations at this time.

It is easy to see why so few humpbacks were sighted in New Zealand waters after the early 1960s. The breeding stocks of whales that once migrated past the country were virtually wiped out by whaling and have yet to show signs of a strong recovery. Ex-whalers, researchers and volunteers, who count whales each year for the Cook Strait Humpback Whale Survey (see page 142), spotted a 'record-breaking' 106 whales during their four-week 2012 count. Another milestone was reached when 23 whales were sighted on 24 June 2012, almost doubling the previous record of 11 whales counted in a single day. Only 59 humpbacks were spotted during the month-long survey in 2013.

The humpback whales that mate and calve around Oceania's South Pacific islands are still thought to be at less than 20 per cent of their original numbers, with a recent population estimate of around 4300 individuals. Eastern Australian humpbacks, on the other hand, have shown better rates of recovery, and their 2012 population is estimated to be around 15,000 individuals. The IUCN has assessed these two groups as a single sub-population, and because of its poor overall recovery has classed it as Endangered.

THREATS AND CONSERVATION

Although humpback whale numbers are slowly recovering, the species still faces threats from global warming, ocean pollution, ship strikes and entanglement in fishing gear. From 2000 to 2012, 20 humpbacks were reported entangled in fishing gear in New Zealand waters. The majority of the whales were tangled in ropes or buoys from craypots, and 14 of the whale entanglements were reported off Kaikoura. In 2003, a Kaikoura man died trying to free a whale with rope wrapped around its tail, and since then the Department of Conservation has devised a programme to free entangled whales that does not require a person to enter the water. Six of the 20 entangled whales were released and two freed themselves; the fate of the others is unknown.

'He is the most gamesome and light-hearted of all the whales, making more gay foam and white water than any other of them.'

Herman Melville, Moby-Dick (1851)

GLOSSARY

abyssal plain Underwater plain consisting mainly of fine sediments and located at around 4000 metres, the average depth of the sea floor. The world's abyssal plains cover almost 150 million square kilometres, close to half the total combined ocean area.

abyssopelagic zone Oceanic zone beginning at 4000 metres and descending to the seabed.

ambergris Greyish waxy substance that forms in the intestines of sperm whales; once widely used in the manufacture of perfumes.

Archaeoceti Ancient whales, including the first Cetacea to enter the sea and all their descendants that do not possess cranial telescoping; gave rise to early modern cetaceans.

Artiodactyla Mammalian order that includes even-toed hoofed mammals such as hippopotamuses, pigs, camels, giraffes, cattle, deer and antelope; the closest relatives of whales among living mammals.

astragalus Ankle bone with two pulley-like joint surfaces, one of which connects to the tibia (shinbone) while the second connects to other ankle bones. The astragalus enhances mobility and is found in all even-toed hoofed mammals (artiodactyls).

auditory bulla Ear bone in whales that houses the inner ear structure.

aurei Curved pins, often made from dolphin teeth, that were fastened to Māori cloaks; important people may have had a number of aurei fastened to their cloak, which would rattle as they walked and signify their approach.

baleen Horny keratinous substance that occurs as a series of comb-like plates hanging side by side in rows suspended from the upper jaw of whales in mysticetes (baleen whales). Fibrous fringes along the inner surface of the plates filter and trap prey (zooplankton and small schooling fish).

baleen whales All whales with baleen plates; they are placed in the infra-order Mysticeti and are known as mysticetes.

bathypelagic zone Oceanic zone that extends from 1000 to 4000 metres in depth.

beak Forward-projecting upper and lower jaws that are prominent in some toothed cetacean species.

beaked whales Toothed cetaceans that are members of the family Ziphiidae, which currently recognises five genera and 21 species.

bioluminescence Light emitted by some marine organisms (as a result of a chemical reaction), used to deter predators or to attract prey or mates.

blowhole External nostril (or nostrils) or respiratory opening of cetaceans. Toothed whales (odontocetes) have one blowhole; baleen whales (mysticetes) have two.

blubber Specialised layer of fatty tissue found between the skin and muscle of all cetaceans. Blubber insulates the body core and stores energy.

bow-riding Action or behaviour of riding on the pressure wave created on the bow of a moving boat or a ship.

bradycardia Condition that occurs when the heart rate is decreased.

breaching When whale or dolphin leaps out of the water; a single leap is called a breach.

bubble-netting Cooperative feeding technique used by humpback whales in which they create 'nets' of bubbles to corral or trap zooplankton or small fish.

by-catch Animals that are not the target species of a fishery but are caught accidentally during fishing operations.

callosities Raised, thickened and rough patches of skin tissue found on a right whale's head; they differ in number, size and configuration, allowing researchers to identify individuals from their unique pattern. Callosities generally harbour large colonies of whale lice (cyamids).

calving interval Period of time from one birth to the next for the females within a population.

caudal peduncle See 'tailstock'.

Cetacea Taxonomic infra-order that includes all extinct and living whales, dolphins and porpoises.

Cetariodactyla Relatively new taxonomic order containing both cetaceans (whales, dolphins and porpoises) and artiodactyls (even-toed ungulates). Cetariodactyla is a hybridised name, combining Cetacea and Artiodactyla.

civavonovono Breastplates worn by chiefs and high-ranking men, often decorated with whale-bone objects.

click Sound produced by a toothed cetacean (odontocete) when it is using echolocation. The sound is usually short (micro-seconds) and 110–150 kilohertz in frequency.

click train Series of rapid clicks that run together so that individual clicks become indistinguishable from one another.

coda Patterned series of clicks, usually comprising 3–20 clicks, used by sperm whales for communication.

copepod Minute crustacean that occurs in great abundance; forms an important role in the marine ecosystem and acts as a food source for some baleen whales as well as other marine creatures.

Cretaceous Geological time period between 144 and 65 million years ago.

Critically Endangered As defined by the IUCN, a taxon that faces an extremely high risk of extinction in the wild.

cyamid Whale louse belonging to the family Cyamidae; usually found colonising the callosities on right whales.

deep scattering layer (DSL) Organisms associated with the edge of light that tend to migrate vertically on a daily or nightly basis, being hundreds of metres below the surface during the day and close to the surface at night.

diatom Single-celled algae (phytoplankton) characterised by a wall of overlapping halves that are impregnated with silica. Diatoms often coat the bodies of whales in polar waters, when they appear as a yellowish sheen.

dimorphism Difference in form (body size, shape or colour) between two individuals or two groups of individuals.

dolphin Cetacean that is usually (but not always) smaller than 4 metres in length and has cone-shaped teeth.

dorsal Upper surface of the back or other body parts.

dorsal fin Fin along the midline of a cetacean's back; found on the majority of species.

drift net Unanchored fishing net that is suspended vertically in the water so that swimming or drifting animals become trapped in its mesh.

echelon feeding Feeding method whereby whales in a group move side by side in a coordinated fashion as they search for prey.

echolocation Production of sound waves, followed by the reception of their echoes, used by cetaceans to locate objects and investigate the surrounding environment.

ecosystem Biological community and its physical environment considered as a whole unit.

eco-tourism Tourism which operates in an environmentally friendly manner.

Endangered As defined by the IUCN, a taxon that is not critically endangered but nonetheless faces a high risk of extinction in the wild in the near future.

endemic Species or race that is restricted to, and only breeds in, a particular locality or region.

Eocene Geological time period from 56 to 34 million years ago.

epipelagic zone Oceanic zone extending from the surface down to 200 metres; also referred to as the sunlit zone.

euphausiids Small shrimp-like crustaceans, members of the family Euphausiidae, usually luminescent, which make up a large proportion of the zooplankton and are commonly referred to as krill.

Exclusive Economic Zone (EEZ) Area extending from a country's coastline to 200 nautical miles offshore in which it has the exclusive sovereign right to manage its resources.

extant Population or species currently in existence, not extinct.

extinct When the last individual from a species or population has died.

factory ship Whaling vessel that hunts, catches and processes whales on board.

filter-feeding Method used by some marine mammals to strain or filter their prey, either through baleen or serrated teeth.

fission–fusion society Free-flowing cetacean group whose members may split apart and come together again, often forming both small and large new groups in the process.

flenching/flensing Process of removing the blubber from a marine mammal carcass.

flipper Flattened and often paddle-shaped forelimbs, also known as pectoral fins, of cetaceans, seals, dugongs and manatees.

flipper-slapping Slapping the pectoral fin or flipper on the surface of the water, often repetitively.

flukes Tail of a cetacean, which is horizontally spread unlike the vertical tail of a fish.

fluking Act of raising the tail or flukes above the water at the beginning of a dive.

foetal folds Grooves in the skin resulting from the curled position of a cetacean foetus; these appear as stripes on newly born calves that may be visible for a few weeks.

food web Feeding cycles (what eats what) in an ecological community; also referred to as a food cycle.

foraging Process of finding, catching and eating food.

genus Group of organisms with shared similarities.

gestation Period of time young are carried in the uterus from conception to birth.

gillnet Anchored net suspended vertically in the water column in which fish or other organisms become entangled and trapped by their gills or other body parts.

global warming Overall rise in the average temperature of Earth's oceans and atmosphere.

Gondwana Ancient supercontinent that included present-day South America, Africa, Arabian Peninsula, Madagascar, India, Australia, New Zealand and Antarctica; fully assembled around 600 million years ago, it started to break up around 160 million years ago.

grasper Whale that grasps its prey with its teeth and then swallows it whole.

grubber Whale (e.g. grey whales) that searches around on the ocean floor for small crustaceans, sea worms and other prey.

gulper Baleen whale that engulfs huge volumes of water containing large masses of prey such as krill or small schooling fish.

ha'akai Ear ornaments, particularly worn by women from the Marquesas, which were fashioned from a single whale or dolphin tooth and inserted through the ear lobe.

hadopelagic zone Oceanic zone taking in the deepest depths in the ocean and extending from 4000 metres to 11,000 metres.

harpoon Fishing spear composed of a barbed or sharp head and a shaft; may be hand-held or attached to a line.

heru Ornamental Māori hair comb used to secure a man's long hair into a topknot.

hoeroa Māori whalebone staff that could be owned and used only by chiefs; sometimes used as a weapon or thrown at a fleeing opponent.

home range Total area covered by individuals in the course of their normal activities, including both feeding and mating/calving.

indigenous whaling Whaling undertaken by groups of people, such as the Inuit, who have traditionally whaled as part of their culture and who claim that their culture continues to depend on whaling for survival or as a way of life.

International Union for the Conservation of Nature and Natural Resources (IUCN) Also known as the World Conservation Union, the IUCN was established in 1948 to encourage a worldwide approach to conservation. Its Red List is an inventory of living things, placing them into categories to denote their worldwide status, from Least Concern to Extinct.

International Whaling Commission (IWC) Initially created by whaling nations in 1946 in order to monitor, regulate and manage whale stocks. It now includes both whaling and non-whaling nations working towards the conservation and management of worldwide whale species, and continues to determine how many, if any, species may be taken from a specific stock each year.

invertebrate Animal without spinal column.

kai Food.

kaimoana Food from the sea.

kaitiaki Guardian.

koha Gift.

koropepe Māori neck ornament coiled in the shape of a manaia-headed (beaked) creature, symbolising an eel in motion; usually made of greenstone or bone.

kotiate Short-handled Māori weapon, made of whale bone or wood, with notches on either side of a broad, flattened blade.

krill Small shrimp-like crustacean zooplankton that are especially abundant in the Antarctic, where they are the main food source of most baleen whale species as well as seal and penguin species.

Least Concern As defined by the IUCN, a taxon that has been evaluated but does not satisfy the criteria to be classified as Near Threatened, Vulnerable, Endangered or Critically Endangered.

Linnaean system Binomial naming system used to classify all living organisms in which each is given two names, a genus name and a species name.

lob-tailing Behaviour in which a whale or dolphin slams its flukes down on the water's surface, often repeatedly; also called tail-slapping.

logging When whales or dolphins breathe slowly and lie quietly on or almost under the surface of the water.

mammary slits Small openings on the underside (ventral side) of female marine mammals that conceal the teats.

marine mammal Mammal that lives in, or depends on, the ocean for its survival. Examples include whales, seals, manatees and polar bears.

marine snow Small particles of decaying animal matter that drift to the ocean floor.

mass stranding Simultaneous stranding of two or more animals, other than a mother–calf pair.

melon Fat-filled region in the 'forehead' of toothed cetaceans, thought to direct and focus sound during echolocation.

Mesonychidae Group of wolf-like animals that lived during the Eocene epoch, once believed to be the primary land ancestors of whales.

mesopelagic zone Oceanic zone beginning at 200 metres and descending to 1000 metres; sometimes referred to as the twilight zone.

migration Seasonal travel from one geographic location to another, often between feeding and breeding grounds.

Mysticeti Cetacean order of whales with baleen, collectively referred to as mysticetes.

Near Threatened As defined by the IUCN, a taxon with more than 1000 mature adults and no current evidence of decline. May include species that are close to qualifying as Vulnerable, but does not include species thought to be under-recorded.

nguru Musical flute, often made from whale teeth, whose shape is unique to New Zealand.

nursery pod Group of whales or dolphins composed of mothers and calves.

ocean trench Narrow, deep depression in the ocean floor; the deepest parts of the ocean are located in the Mariana Trench (11,000 metres) and the Kermadec Trench (over 10,000 metres).

Odontoceti Cetacean order containing the toothed whale, dolphin and porpoise species, collectively known as odontocetes.

Pacific Oceanscape Launched in 2011 by 15 South Pacific nations, this is one of the largest government-endorsed marine management initiatives, covering 38.5 million square kilometres of ocean.

patu parāoa Short-handled Māori weapon fashioned from the jawbone of a sperm whale (parāoa means both 'sperm whale' and 'chief').

pelagic Living or occurring in the open sea.

pelagic whaling Whaling operation that does not depend on a land station for processing whales but instead uses a floating factory ship.

peue ei/peue koi'o Headdress known as a porpoise-toothed crown, primarily made in the Marquesas from dolphin teeth and glass beads strung together on a coconut-fibre cord.

photo identification Use of photographic images to identify individuals within a population.

photophore Light-emitting organ composed of a cluster of light-producing cells, a reflector and a lens found on the bodies of luminescent organisms.

phytoplankton Plant plankton.

picoplankton One of the tiniest phytoplankton species, smaller than the width of a hair.

plankton Passively drifting or weakly swimming plant or animal organisms that occur in swarms near the surface of the water; the base of the oceanic food web.

porpoise Small cetaceans in the family Phocoenidae, all of which have spade-shaped teeth.

porpoising Behaviour exhibited by cetaceans, penguins and seals, where a rapidly swimming animal partially leaps out of the water repeatedly.

purse seine Type of fishing gear that surrounds fish schools with a vertical curtain of netting, which is then pursed, or closed off, at the bottom in order to prevent fish from escaping.

rei puta Ornament made from a split sperm whale tooth, which has carved eyes and is often worn as a necklace by high-ranking men. Rei puta were made by many South Pacific island cultures, including Māori.

rorqual Any of the species of baleen whales with throat grooves (ventral pleats) that expand during feeding to increase the capacity of the mouth.

scrimshaw Artform in which whale bones or teeth are polished and then etched with a sharp object such as a knife, needle or nail. The etched picture is then darkened with ash or other material such as boot black.

seamount Steep-sided circular or elliptical feature that rises from the sea floor to an elevation reaching or exceeding 1000 metres.

sexual dimorphism When males and females of the same species differ in size (such as the male and female sperm whale), or when one sex has distinctive secondary sexual characteristics (such as the male narwhal tusk).

sisi Fijian necklace made from whole whale teeth, worn only by chiefs or influential men.

site fidelity Tendency to return repeatedly to the same site to mate or give birth every breeding and calving season.

skimmer/skim feeder Baleen whale that slowly swims along the surface of the water, or just under it, with its mouth open continuously, filtering the water and trapping prey in its baleen as it does so.

solo dolphin Dolphin (or whale) that forsakes its own kind and seeks out the companionship of a species that is not its own, usually humans.

South Pacific Whale Research Consortium (SPWRC) Group formed in 2001 by independent scientists conducting non-lethal research on humpback and other whale species in South Pacific waters.

Southern Ocean Whale Sanctuary
Created by the IWC in 1994, the sanctuary covers 50 million square kilometres of ocean and excludes all commercial whaling

species Group of interbreeding populations of animals that are reproductively isolated from other such populations.

spermaceti Distinctive form of oil composed of wax esters and triglycerides found in the spermaceti organ in the head of sperm whales.

spermaceti organ Elongated barrel-shaped structure making up much of a sperm whale's head and comprising soft, spongy tissue filled with spermaceti oil.

spout Exhaled breath of a cetacean, made up of condensed water vapour mixed with sea water from around the blowhole.

spyhopping Behaviour whereby a cetacean raises its head out of the water (presumably to look above the surface) and then lowers it back in, usually with very little splash.

stock Population of whales found in a specific area of the world, e.g. the New Zealand stock of southern right whales.

stranding When a marine animal comes ashore (beaches itself) or is cast ashore.

sub-adult Individual that is older than an infant and younger than an adult; often used in age/class studies as an additional category – infant, juvenile, sub-adult and adult.

subantarctic islands Islands located in a transitional zone between temperate waters and the polar waters off Antarctica; New Zealand's antarctic islands include The Snares/Tini Heke, Bounty Islands, Antipodes Islands, Auckland Islands and Campbell Islands.

submarine canyons Steep underwater chasms that are gouged out by turbidity currents containing a mix of sea water and sediments.

subsistence hunting Hunting whales (and other animals) for survival as opposed to commercial gain.

subspecies Population that is geographically isolated from other populations of the same species and is evolving in its own direction.

suction feeding Capture of prey using suction, generally with the tongue employed as a type of piston to create vacuum pressure.

tabua Single ceremonial sperm whale tooth, given as a gift or used to barter or trade; often attached to a cord but never worn.

taiaha Long-handled weapon, used by Māori warriors somewhat like a sword to feint, thrust and parry. The top of the weapon is often intricately carved.

tail flukes Two horizontally flattened structures that together form the tail of a whale.

tail-slapping See 'lob-tailing'.

tailstock Area of a whale's body in front of the tail flukes; also known as the caudal peduncle.

taniwha Spiritual creatures that live in oceans, inland waters, dens or dark caves. Taniwha may protect or guard the area in which they live, while others may travel with humans in order to protect them during journeys. Taniwha may take on many forms, including that of whales.

taonga Treasure.

taxonomy Science of classifying organisms into species, genera and so on to higher categories.

tetrapod Creature with four limbs thought to have evolved from lobe-finned fish species that lived in shallow, muddy waterways.

thermoregulation Process of controlling and maintaining a constant body temperature in an animal as the outside temperature changes. Examples in human are shivering to warm up or sweating to cool down.

throat pleats Folds of skin in the throat of some baleen whale species that expand when it is feeding, allowing it to take in huge mouthfuls of water and prey; also called ventral grooves.

tohorā Māori/South Pacific term for whales; sometimes used to denote the right whale.

toothed whales Order of whales known as Odontoceti.

trypot Large iron pot that was placed over a fire in order to boil down blubber from whales or seals.

tumutumu Māori musical instrument, often made from whale bone, which is struck in different places to produce sounds varying in pitch and intensity.

tusk Modified tooth. Narwhal males have one long tusk, while many male beaked whale species have two to four tusks.

ungulate Any animal belonging to the taxonomic group comprising the hooved animals and their derivatives, such as Artiodactyla, Cetacea, Sirenia etc.; includes extinct species and other fossil groups.

upwelling Area of the ocean where warm currents push cooler waters to the surface, bringing with them a large quantity of nutrients.

vagrant Animal that has moved outside of the usual limits of distribution for its species or population.

ventral Underside or bottom of an animal.

ventral groove See 'throat pleats'.

Vulnerable As defined by the IUCN, a taxon that faces a high risk of extinction in the medium-term future.

wahaika Single-blade weapon used by Māori, distinguished by having a concave recess in the blade that often contains a carved human figure.

wake-riding Behaviour of riding or surfing in the wake created by a boat or ship as it travels through the water.

wāseisei Fijian necklace made from split whale teeth, and worn only by chiefs and other men of influence.

whale Member of the mammalian infra-order Cetacea. The term is generally applied to the larger baleen species, along with some toothed species, but inconsistently used for cetaceans of different sizes; it is also used in the common names for some members of the Delphinidae family, such as pilot whales and killer whales.

whalebone Term used to describe baleen, especially during the whaling years, when the material was used in the manufacture of corsets, upholstery, umbrellas etc.

Zealandia New Zealand continent, now mostly submerged, stretching from New Caledonia to Campbell Islands, which separated from Gondwana around 100–80 million years ago.

zooplankton Animal forms of plankton comprising numerous species, many of which (krill, copepods, molluscs etc.) are fed upon by baleen whales.

BIBLIOGRAPHY

Adams, Douglas. (1979). *The Hitchhiker's Guide to the Galaxy*. London: Pan Books, Ltd. Quote from p.119.

Aldrich, H.L. (1889). *Arctic Alaska and Siberia, or, Eight Months with the Arctic Whalemen*. Chicago and New York: Rand, McNally and Company. Quote from p. 34.

Allison, C. (2010). IWC Summary Catch Database Version 5.0, 1 October 2010.

American Museum of Natural History. (2006). *Ocean: The world's last wilderness revealed*. New York, London: DK Publishing Ltd. 512 pp.

Anderson, Iain. (2007). *The Surface of the Sea: Encounters with New Zealand's upper ocean life*. Auckland: Reed Publishing. 151 pp.

Anon. (1956). 'Opo Will Soon Have Full Protection'. *Weekly News* (Auckland) 15 February. Cover.

Arrian. (1983). *History of Alexander and Indica*. Loeb Classical Library. Cambridge Mass: Harvard University Press. 589 pp.

Baker, Alan N. (1978). The status of Hector's dolphins (van Beneden) in New Zealand waters. Report to the International Whaling Commission 28: 331–34.

Baker, Alan. (1999). *Whales and Dolphins of New Zealand and Australia*. Wellington: Victoria University Press. 133 pp.

Baker, Alan N. and Madon, Bénédicte. (2007). Bryde's whales (*Balaenoptera* cf. *brydei* Olsen 1913) in the Hauraki Gulf and northeastern New Zealand waters. *Science for Conservation* 272. Wellington: Department of Conservation. 23 pp.

Baker, Alan N. and van Helden, Anton L. (1990). First record of the dwarf sperm whale, *Kogia simus* (Owen), from New Zealand. National Museum of New Zealand Records 3(12): 125–30.

Baker, Alan, Smith, Adam N.H. and Pichler, Franz B. (2002). Geographical variation in Hector's dolphin: Recognition of new sub-species of *Cephalorhynchus hectori*. *Journal of the Royal Society of New Zealand* 32(4): 713–27.

Baker, C.S., Chilvers, B.L., Constantine, R., DuFresne, S., Mattlin, R.H., van Helden, A. and Hitchmough, R. (2010). Conservation status of New Zealand marine mammals (suborders Cetacea and Pinnipedia). *New Zealand Journal of Marine and Freshwater Research* 44(2): 101–15.

Batson, Peter. (2003). *Deep New Zealand: Blue water, black abyss*. Christchurch: Canterbury University Press. 240 pp.

Beale, Thomas. (1973). *The Natural History of the Sperm Whale*. London: The Holland Press, London. Quote from p. 183.

Beston, Henry. (1928). *The Outermost House: A year of life on the great beach of Cape Cod*. New York: Doubleday and Doran. Quote from p. 24.

Bullen, Frank T. (1923). *The Cruise of the 'Cachalot'*. London: John Murray. 188 pp. [First published in 1899, London: Smith Elder.]

Byatt, Andrew, Fothergill, Alastair and Holmes, Martha. (2001). *The Blue Planet: A natural history of the oceans*. London: BBC Worldwide Ltd. 384 pp.

Carroll, E.L., Patenaude, N., Alexander, A., Steel, D., Harcourt, R., Childerhouse, S., Smith, S., Bannister, J., Constantine, R. and Baker, C.S. (2011). Population structure and individual movement of southern right whales around New Zealand and Australia. *Marine Ecology Progress Series* 432: 257–68.

Carroll, E.L., Patenaude, N.J., Childerhouse, S.J., Kraus, S.D., Fewster, R.M. and Baker, C.S. (2011). Abundance of the New Zealand subantarctic southern right whale population estimated from photo-identification and genotype mark-recapture. *Marine Biology* 158(11): 2565–75.

Carroll, Emma L., Rayment, William R., Alexander, Alana M., Baker, C. Scott, Patenaude, Nathalie J., Steel, Debbie, Constantine, Rochelle, Cole, Rosalind, Boren, Laura J. and Childerhouse, Simon. (2013). Reestablishment of former wintering grounds by New Zealand southern right whales. *Marine Mammal Science*, doi: 10.1111/mms.12031.

Carwardine, Mark. (1995). *Whales, Dolphins and Porpoises*. Sydney: Harper Collins Australia. 256 pp.

Carwardine, Mark. (2006). *Collins Wild Guide: Whales and dolphins*. London: Harper Collins. 256 pp.

Chambers, John H. (2004). *A Traveller's History of New Zealand and the South Pacific Islands*. London: Windrush Press. 314 pp.

Christensen, Line Bang. (2006). Marine mammal populations: Reconstructing historical abundances at the global scale. Fisheries Centre Research Report 14(9): 1–161. Quote from p. 83.

Clapham, Phillip. (1997). *Whales*. Moray (Scotland): Colin Baxter Photography. 132 pp.

Clapham, P., Powell, J., Reeves, R. and Stewart, B. (2002). *National Audubon Society Guide to the Marine Mammals of the World*. New York: Alfred Knopf, Inc. 528 pp.

Committee on Taxonomy. (2012). List of marine mammal species and subspecies. Society for Marine Mammology, www.marinemammalscience.org, cited on 6 November 2013.

Constantine, R., Brunton, D.H. and Baker, C.S. (2003). The behavioural ecology and effects of tourism on bottlenose dolphins of northeastern New Zealand. *Science for Conservation* 153. Wellington: Department of Conservation. 26 pp.

Constantine, Rochelle, Jackson, Jennifer A., Steel, Debbie, Baker, C. Scott, Brooks, Lyndon, Burns, Daniel, Clapham, Phillip, Hauser, Nan, Madan, Bénédicte, Mattila, David, Oremus, Marc, Poole, Michael, Robbins, Jooke, Thompson, Kirsten and Garrigue, Claire. (2012). Abundance of humpback whales in Oceania using photo-identification and microsatellite genotyping. *Marine Ecology Progress Series* 453: 249–61.

Currey, Rohan J.C., Dawson, Stephen M. and Slooten, Elisabeth. (2009). An approach for regional threat assessment under the IUCN Red List criteria that is robust to uncertainty: The Fiordland bottlenose dolphins are critically endangered. *Biological Conservation* 142: 1570–79.

Dakin, W.J. (1963). *Whalemen Adventurers*. Sydney (Australia): Sirius Books. 285 pp.

Dalley, Bronwyn and McLean, Gavin (eds). (2005). *Frontier of Dreams: The Story of New Zealand*. Auckland: Hodder Moa, Hachette Livre. 416 pp.

Darwin, Charles. (1859). *On the Origin of Species by Means of Natural Selection*. London: John Murray Publisher. 502 pp.

Davidson, Janet. (1987). *The Prehistory of New Zealand*. Auckland: Longman Paul Ltd. 270 pp.

Dawbin, W.H. (1954). Maori whaling. *The Norwegian Whaling Gazette* 8: 433–45.

Dawbin, W.H. (1956). The migrations of humpback whales which pass the New Zealand coast. Royal Society New Zealand 84: 147–96.

Dawson, S.M. and Slooten, E. (1993). Conservation of Hector's dolphins: The case and process which led to the establishment of the Banks Peninsula Marine Mammal Sanctuary. *Aquatic Conservation: Marine and Freshwater Ecosystems* 1: 207–21.

Dawson, S.M. and Slooten, E. (1994). Hector's dolphins, *Cephalorhynchus, hectori* (van Beneden, 1881). In S.H. Ridgway R. and Harrison (eds). *Handbook of Marine Mammals*. Volume 5. London: Academic Press. XX pp.

Dawson, S.M. and Slooten, E. (1996). *Down-under Dolphins*. Christchurch: Canterbury University Press. 60 pp.

de Lacépède, Bernard Germain. (1804). *Histoire naturelle des Cétacés*. Paris: Plassan. XLIV, 329 S. : 16 ill.

Department of Conservation. (2003). Humpback whales around New Zealand. Conservation Advisory Science Notes 287. Wellington: Department of Conservation. 35 pp.

Dieffenbach, E. (1843). *Travels in New Zealand*. Volumes 1 and 2. London: John Murray. 431 pp

Doak, Wade. (1981). *Dolphin, Dolphin*. Auckland: Hodder and Stoughton. 245 pp.

Doak, Wade. (1988). *Encounters with Whales and Dolphins*. Auckland: Hodder and Stoughton. 250 pp.

Doak, Wade. (1995). *Friends in the Sea: Solo dolphins in New Zealand and Australia*. Auckland: Hodder Moa Beckett. 144 pp.

Eiseley, Loren. (1959). *The Immense Journey*. New York: Vintage Books, Random House. 224 pp.

Forster, George. (2000). *A Voyage Round the World*. Volume 1. Nicholas Thomas and Oliver Berghof (eds). Honolulu: University of Hawaii Press. xlvii, 475 pp

Flintoff, Brian. (2011). *Kura Koiwi: Bone Treasures*. Nelson: Craig Potton Publishing. 157 pp. Quote from p. 75.

Fordyce, R. Ewan, Mattlin, Robert H. and Dixon, Joan M. (1984). Second record of spectacled porpoise from subantarctic southwest Pacific. Scientific Reports of the Whale Research Institute, Tokyo 35: 159–64.

Fordyce, R. Ewan and Marx, Felix G. (2013). The pygmy right whale *Caperea marginata*: The last of the cetotheres. Proceedings of the Royal Society B 280(1753). doi: 10.1098/rspb.2012.2645.

Fowles, John. (1984). 'The Blinded Eye.' Second Nature. London: Jonathan Cape.

Garland, Ellen C., Goldizen, Anne W., Rekdahl, Melinda L., Constantine, Rochelle, Garrigue, Claire, Hauser, Nan Deechler, Poole, M. Michael, Robins, Jooke and Noad, Michael J. (2011). Dynamic horizontal cultural transmission of humpback whale song at the ocean basin scale. *Current Biology* 21(8): 687–91.

Gaskin, D.E. (1964). Return of the southern right whale (*Eubalaena australis* Desm.) to New Zealand waters. *Tuatara* 12(2): 115–18. Quotes from pp. 115, 20.

Gaskin, D.E. (1965). New Zealand whaling and whale research, 1962–64. *New Zealand Science Review* 23(2): 19–22.

Gaskin, D.E. (1968). Composition of schools of sperm whales, *Pyseter catadon. New Zealand Journal of Marine and Freshwater Research* 3: 480.

Gill, Peter and Burke, Cecilia. (2011). *Whale Watching: Australian and New Zealand waters*. 3rd ed. Sydney: New Holland Publishers (Australia). 148 pp.

Gordon, Dennis P. (ed.). (2010). *New Zealand Inventory of Biodiversity: Kingdom Animalia*. Volume 1. Christchurch: Canterbury University Press. 566 pp.

Grady, Don. (1982). *The Perano Whalers*. Wellington: Reed. 238 pp.

Grady, Don. (1986). *Sealers and Whalers in New Zealand Waters*. Auckland: Reed Methuen. 307 pp.

Hamner, Rebecca M., Constantine, Rochelle, Oremus, Marc, Stanley, Martin, Brown, Phillip and Baker, C. Scott. (2013). Long-range movement by Hector's dolphins provides potential genetic enhancement for critically endangered Maui's dolphin. *Marine Mammal Science*. doi:10.1111/mms.12026.

Harris, Jan. (1994). *Tohora: The Story of Fyffe House*, Kaikoura. Wellington: New Zealand Historic Places Trust. 56 pp.

Heberley, Heather. (2002). *Last of the Whalers: Charlie Heberley's story*. Auckland: Cape Catley Ltd. 212 pp.

Horsman, Paul. (2005). *Out of the Blue*. London: New Holland Publishers (UK) Ltd. 160 pp.

Howe, K.R. (ed.). (2006). *Vaka Moana: Voyages of the ancestors – The discovery and settlement of the Pacific*. Auckland: Auckland War Memorial Museum and David Bateman. 360 pp.

Hutchinson, Stephen and Hawkins, Lawrence E. (2004). *Oceans: The Macmillan visual guide*. Sydney: Pan Macmillan. 303 pp.

International Whaling Commission (2009). Southern Hemisphere Catch Totals: 1900–2005. In William F. Perrin, J.G.M. Würsig, and Bernd Thewissen (eds) *Encyclopedia of Marine Mammals*. 2nd ed. Amsterdam: Elsevier, Academic Press: 1240

Jefferson, Thomas A., Pitman, Robert L. and Webber, Marc A. (2008). *Marine Mammals of the World*. Amsterdam: Elsevier, Academic Press. 592 pp.

Jensen, F.H., Perez, Jacobo Marrero, Johnson, Mark, Soto, Natacha Aguilar and Madsen, Peter T. (2011). Calling under pressure: Short-finned pilot whales make social calls during deep foraging dives. Proceedings of the Royal Society B. doi: 10.1098/rspb.2010.2604.

Kerr, I.S. (1976). *Campbell Island: A history*. Wellington: Reed. 182 pp.

Lawrence, D.H. (1932). 'Whales weep not'. In Richard Aldington and Giuseppe Orioli (eds). *Last Poems*. Florence: Giuseppe Orioli. 235 pp.

Lee-Johnson, Eric and Lee-Johnson, Elizabeth. (1994). *Opo: The Hokianga dolphin*. Auckland: David Ling Publishing. 47 pp

Martin, Stephen. (2001). *The Whales' Journey*. Crow's Nest, NSW (Australia): Allen and Unwin. 251 pp.

McKenna, Virginia. (1992). *Into the Blue*. New York: Harper Collins. 144 pp.

McNab, Robert. (1975). *Old Whaling Days*. Auckland: Golden Press. [First published in 1913, Christchurch: Whitcombe and Tombs]. Quotes from pp. 9–10.

Melville, Herman. (1851). *Moby Dick*. Secaucus, N.J.: Longriver Press [1976 edition]. Quotes from pp. 376, 457–59, 131.

Merriman, Monika G., Markowitz, Tim M., Harlin-Cognato, April D. and Stockin, Karen A. (2009). Bottlenose dolphin (*Tursiops truncatus*) abundance, site fidelity, and group dynamics in the Marlborough Sounds, New Zealand. Aquatic Mammals 35(4): 511–22.

Morton, Harry. (1982). *The Whale's Wake*. Dunedin: University of Otago Press. 396 pp.

Morzer Bruyns, W.F.J. and Baker, A.N. (1972). Notes on Hector's dolphins, *Cephalorhynchus hectori* (van Beneden) from New Zealand. Records of the Dominion Museum 8(9): 125–37.

Mowat, Farley. (2004). *Sea of Slaughter*. Mechanicsburg, PA: Stackpole Books. 420 pp. Quote from p. 199.

Mulcahy, Kate and Peart, Raewyn. (2012). *Wonders of the Sea: The protection of New Zealand's marine mammals*. Auckland: Environmental Defence Society. 320 pp.

New Zealand Ministry for Culture and Heritage. *Maori Peoples of New Zealand Nga iwi o Aotearoa. Te Ara, The Encyclopedia of New Zealand*. Auckland: David Bateman Ltd. 294 pp.

Ommanney, F.D. (1971). *Lost Leviathan*. London: Hutchinson. 280 pp.

Orbell, Margaret. (1995). *The Illustrated Encyclopedia of Māori Myth and Legend*. Christchurch: Canterbury University Press. 274 pp. Quote from p. 233.

Patenaude, Nathalie J. (2003). Sightings of southern right whales around 'mainland' New Zealand. *Science for Conservation* 225. Wellington: Department of Conservation. 15 pp.

Payne, Roger. (1995). *Among Whales*. New York: Charles Scribner's Sons. 431 pp.

Perrin, W.F., Würsig, B. and Thewissen, J.G.M. (eds). (2009). *Encyclopedia of Marine Mammals*. 2nd ed. Amsterdam: Elsevier, Academic Press. 1352 pp.

Pitman, R.L., Durban, J.W., Greenfelder, M., Guinet, C., Jorgensen, M., Olson, P., Plana, J., Tixier, P. and Towers, J.R. (2011). Observations of a distinctive morphotype of killer whale (*Orcinus orca*), type D, from the subantarctic waters. *Polar Biology* 34: 303–06.

Pitman, Robert L. and Ensor, Paul. (2003). Three forms of killer whales (*Orcinus orca*) in Antarctic Waters. *Journal of Cetacea Research and Management* 5(2): 131–39.

Pliny the Elder, (AD 79). *Historia Naturalis*, 1X, ii, 4. [Philemon Holland's translation, 1604.]

Prickett, Nigel. (2002). *The Archaeology of New Zealand Shore Whaling*. Wellington: Department of Conservation. 151 pp.

Rayment, W., Davidson, A., Dawson, S, Slooten, E. and Webster, T. (2012). Distribution of southern right whales on the Auckland Islands calving grounds. *New Zealand Journal of Marine and Freshwater Research* 46(3): 431–36.

Richards, Rhys. (1982). *Whaling and Sealing at the Chatham Islands*. Roebuck Society Publication 21. Canberra, Australia: Roebuck Society.

Robertson, R.B. (1956). *Of Whales and Men*. Suffolk: Richard Clay and Co. 247 pp.

Robson, Frank. (1984). *Strandings: Ways to save whales*. Johannesburg: The Science Press. 124 pp.

Robson, Frank. (1988). *Pictures in the Dolphin Mind*. Auckland: Reed Methuen. 135 pp.

Russell, Kirsty. (2001). The North Island Hector's dolphin is vulnerable to extinction. *Marine Mammal Science* 17(2): 366–71.

Scheffer, Victor B. (1969). *The Year of the Whale*. New York: Charles Scribner's Sons. 213 pp. Quote from p. 5.

Shirihai, Hadoram and Jarrett, Brett. (1996). *Whales, Dolphins and Seals: A field guide to the marine mammals of the world*. London: A and C Black. 384 pp.

Smolker, Rachel. (2001). *To Touch a Wild Dolphin*. New York: Nan A. Talese. 188 pp.

Stace, Glenys. (2005). *Blue New Zealand*. Auckland: Penguin Group New Zealand. 32 pp.

Stockin, Karen A., Amaral, Ana R., Latimer, Julie, Lambert, David M. and Natoli, Ada. (2013). Population genetic structure and taxonomy of the common dolphin (*Delphinus* sp.) at its southernmost range limit: New Zealand waters. *Marine Mammal Science*. doi: 10.1111/mms.12027.

Stockin, Karen A. and Orams, Mark B. (2008). The status of common dolphins (*Delphinus delphis*) in New Zealand waters. *Journal of Cetacean Research and Management*. http://iwcoffice.co.uk/_documents/sci_com/SC61docs/SC-61-SM20.pdf.

Tezanos-Pinto, Gabriela and Constantine, Rochelle. (2013). Decline in local abundance of bottlenose dolphins (*Tursiops truncatus*) in the Bay of Islands, New Zealand. *Marine Mammal Science*. doi: 10.1111/mms.12008.

Torres, L.G., (2013). Evidence for an unrecognized blue whale foraging ground in New Zealand. *New Zealand Journal of Marine and Freshwater Research*, 47(2): 235–48.

Townsend, Charles Haskins. (1935). The distribution of certain whales as shown by the logbook records of American whaleships. *Zoologica* 19(1–2): 1–50.

Van Waerebeek, K., Leaper, R., Baker, A.N., Papastavrous, V., Thiele, D., Dindaly, K., Donovan, G. and Ensor, P. (2010). Odontocetes of the Southern Ocean Sanctuary. *Journal of Cetacean Research and Management* 11: 315–46.

Visser, Ingrid N. (2005). *Swimming with Orca: My life with New Zealand's killer whales*. Auckland: Penguin Books. 204 pp.

Visser, Ingrid N., Zaeschmar, Jochen, Halliday, Jo, Abraham, Annie, Ball, Phil, Bradley, Robert, Daly, Shamus, Hatwell, Tommy, Johnson, Tammy, Johnson, Warren, Kay, Laura and Maessen, Tim. (2010). First record of predation on false killer whales (*Pseudorca crassidens*) by killer whales (*Orcinus orca*). *Aquatic Mammals* 36(2): 195–204.

Wakefield, Edward Jerningham. (1845). *Adventure in New Zealand, from 1839 to 1844: With some account of the beginning of the British colonization of the islands*. Volume 1. London: John Murray. 482 pp.

Whitaker, I. (1985). The Kings Mirror (Konung's skuggsjá). *Polar Records* 22: 615–27.

Wiseman, Nicky, Parsons, Stuart, Stockin, Karen A. and Baker, C. Scott. (2011). Seasonal occurrence and distribution of Bryde's whales in the Hauraki Gulf. *Marine Mammal Science*. doi: 10.1111/j.1748-7692.2010.00454.x.

Würsig, Bernd and Würsig, Melany (eds). (2010). *The Dusky Dolphin: Master acrobat of different shores*. Amsterdam: Academic Press, Elsevier Ltd. 416 pp.

Würtz, Maurizio and Repetto, Nadia. (2009). *Dolphins and Whales*. Novara, Italy: White Star Publishers. 168 pp.

Yablokov, A.V., Zlemsky, V.A., Mikhalev, Y.A, Tormosov, D.D. and Berzin, A.A. (1998). Data on Soviet whaling in the Antarctic in 1947–1972 (population aspects). *Russian Journal of Ecology* 29(1): 43–48. Quote from p. 45.

IMAGE CREDITS

All photographs are by Barbara Todd and all illustrations are by Geoff Cox, unless credited below.

FRONT MATTER

vii Kim Westerskov; **ix** Brian Flintoff; **x** Kim Westerskov

ALL ABOUT WHALES

5 Tom Van Sant/Geosphere Project, Santa Monica/Science Photo Library; **6–7** all images Ron Blakey, Colorado Plateau Geosystems, Inc; **8 top** Kim Westerskov; **bottom** Getty Images; **9** John Durham/Science Photo Library; **11** Kim Westerskov; **13 bottom** Norman Heke, Te Papa; **15** Fairfax Media/*Dominion Post*; **17 left** Jeremy Glyde, Te Papa, based on a map by NIWA; **right** NIWA; **19** Dennis Buurman; **23 bottom** Richard Barnes/OTTO/Raven & Snow; **24 bottom** Richard Barnes/OTTO/Raven & Snow; **25 top** Jeremy Glyde, Te Papa, based on a map by Ron Blakey, Colorado Plateau Geosystems, Inc; **26** Te Papa; **27–28** Jeremy Glyde, Te Papa, based on casts by Research Casting International; **29** Kim Westerskov; **30** Brandon Cole; **31** all photos Ewan Fordyce; **32 top** Michael Hall, Te Papa; **bottom left** Robert W. Boessenecker, Fordyce Research Group; **bottom right** Donald E. Hurlbert; **33** Ewan Fordyce; **34** Mark Jones; **37 top** John K.B. Ford/Ursus/SeaPics.com; **41 right** Alexander Turnbull Library, MNZ-0065-1/4-F; **42 left** Thewissen-lab, NEOMED; **43 top** Kim Westerskov; **45 bottom right** Mark Jones; **48 top** Armin Maywald/SeaPics.com; **bottom** Mark Johnson; **49 top** Marine Themes; **50** Steve Dawson, from Liz Slooten & Steve Dawson, *Dolphins Down Under*, Otago University Press, 2013; **51** Marine Themes; **54** Jeremy Glyde, Te Papa; **55** Kim Westerskov; **57 top** Marine Themes; **59 bottom** Dennis Buurman; **60 bottom** Kim Westerskov; **61 top** ardea.com; **bottom** Dennis Buurman; **62 top** Emma Newcombe; **bottom** Kim Westerskov; **68** Department of Conservation; **70 top** Roger Sutherland; **72** Jeremy Glyde, Te Papa; **74 top** Alexander Turnbull Library, PAColl-6208-31; **centre** Horatio Gordon Robley, *Pataka*, collection of Te Papa; **bottom** unknown carver, attributed to Ngāi Tūhoe, collection of Te Papa; **75** Alexander Turnbull Library, E-296-q-025-1; **76 top left** Brian Flintoff; **top right** Department of Conservation; **bottom** Michael Hall, Te Papa; **77** Ross D. Wearing, ProMotion Media; **78 top and bottom left** Michael Hall, Te Papa; **bottom right** Department of Conservation

OF WHALES AND MAN

80–81 Sydney Parkinson, *Representation of a war canoe of New Zealand, with a view of Gable End Foreland, 1769–70*. Alexander Turnbull Library, B-085-013; **82** dolphin-painted fresco from the Knossos Palace, Greece, 16th Century BC, De AGostini/The British Library Board, 86034280; **83** Science Photo Library; **84** The Trustees of the British Museum; **85 top** Bibliothèque Nationale de France; **bottom** maker unknown, collection of Te Papa; **86 top** AFP Photo; **bottom** American Museum of Natural History Library, 4587; **87** Auckland Art Gallery Toi o Tāmaki, gift of Messrs Samuel Vaile & Sons, 1914; **88 top** Jeremy Glyde, Te Papa; **bottom** Fiji Museum; **89** all objects maker unknown, collection of Te Papa; **90** Michel Tuffery, collection of Te Papa; **91** Norman Clark; **93 top** Michael Hall, Te Papa; **bottom** Robyn Kahukiwa/Penguin Books; **94** both photos Michael Hall, Te Papa; **95** Michael Hall, Te Papa; **96 top** Museum of Wellington City and Sea; **bottom** Daniel Murray; **97** Michael Hall, Te Papa; **98** Michael Hall, Te Papa; **100 left** Richard Dresser; **right** Michael Hall, Te Papa; **101 left** Ramari Stewart; **top right** Shane Cross, Department of Conservation; **bottom** maker unknown, collection of Te Papa; **102** Leslie Adkin, collection of Te Papa; **103 top left** carver unknown, collection of Te Papa; **top centre** carver unknown, attributed to Te Rūnanga a Iwi o Ngāpuhi, collection of Te Papa; **top right** carver unknown, collection of Te Papa; **centre right** carver unknown, collection of Te Papa; **bottom** Alexander Turnbull Library, PUBL-0095-3-453; **104** James McDonald, collection of Te Papa; **105** all objects carver unknown, collection of Te Papa; **second from right** attributed to Te Āti Awa; **106** Brian Flintoff; **107 top left** carver unknown, attributed to Te Rūnanga a Iwi o Ngāpuhi, collection of Te Papa; **top centre and right** carver unknown, collection of Te Papa; **bottom** Brian Flintoff; **108** Grant Dixon, Hedgehog House; **110 top** collection of Te Papa; **bottom** New Bedford Whaling Museum; **111** Alexander Turnbull Library, A-032 026; **112** Whanganui Regional Museum collection; **113 top** Alexander Turnbull Library, A-191-009; **bottom** collection of Te Papa; **114 top** *The Family Doctor*; **bottom** Alexander Turnbull Library, 10X8-1012-G; **115 top** Sir George Grey Special Collections, Auckland Libraries, AWNS-19131016-44-5; **bottom** Alexander Turnbull Library A-032-025; **116** Te Papa; **117 right** Alexander Turnbull Library, PAColl-5800-12; **118 top left** Alexander Turnbull Library, 1/2-022935-F; **bottom left** Alexander Turnbull Library,

PUBL-0020-05-3; **right** National Library of New Zealand; **119** Tairāwhiti Museum; **inset** Alexander Turnbull Library, 1/2 027077-G; **120 left** Christchurch City Libraries, CCL PhotoCD 7, IMG0071; **right** Alexander Turnbull Library, G-450; **121 left** National Library of New Zealand; **right** Alexander Turnbull Library, 1/2-052156-F; **122 top** Sir George Grey Special Collections, Auckland Libraries, AWNS 19010726-6-2; **bottom** Picton Historical Society, Inc; **123 top** Alexander Turnbull Library, 1/2-051326-F; **bottom** *Fairfax Media/Manawatu Evening Standard*; **124 left** Te Papa; **right** Alexander Turnbull Library, PAColl-8880; **125 top** Fairfax Media; **bottom left** Alexander Turnbull Library, WA-25237-G; **bottom right** Hocken Collections Uare Taoka o Hakena, P11-007/1; **126** Sir George Grey Special Collections, Auckland Libraries, AWNS 19081224-11-3; **127 top** maker unknown, collection of Te Papa; **bottom** Alexander Turnbull Library, EP/1958/1125-F; **128** Les Stone, Greenpeace; **129** I.P. Golovlev; **131** Kate Davison, Greenpeace; **132 top and bottom left** Heberley collection; **bottom right** Department of Conservation; **133** Heberley collection; **background** Sourced from LINZ, NZMS 25 Map of the North Island, Crown Copyright Reserved; **135** CCI Archives/Science Photo Library; **137 top** Sir George Grey Special Collections, Auckland Libraries, AWNS-19371110-53-3; **bottom** Fairfax Media/*The Evening Post*; **139 top** Nan Hauser; **centre** Keiko Sekiguchi; **141** Alexander Turnbull Library, 1/2-004109-F; **142 top** Sir George Grey Special Collections, Auckland Libraries, AWNS 19390719-48-1; **bottom** Nadine Bott; **144 top** Dennis Buurman; **147 top** Alexander Turnbull Library, 1/2-016340-G; **148 top** Diana Baker; **bottom** Alexander Turnbull Library, EP/1981/1170/15-F; **149 left** Warwick Wilson, Te Papa; **right** Jody Weir; **150** Paul Hilton, Greenpeace; **151** Rochelle Constantine; **152** Kim Westerskov; **153** Jeremy Sutton-Hibbert, Greenpeace; **154** Getty Images; **156** Project Jonah; **157 bottom** maker unknown, collection of Te Papa; **158 top** Pacific Whale Foundation; **bottom** Jeremy Glyde, Te Papa, based on information from Conservation International; **159** Jeremy Glyde, Te Papa; **160** Department of Conservation; **161** Kim Westerskov; **163 top** Kim Westerskov; **bottom** Alex Brandon, AP Images; **164 top** Kim Westerskov; **bottom** James Sutherland; **165 top** Meika Surgenor; **166** all photos Trudi Scott; **167** Glenn Lockitch, Sea Shepherd Australia; **168** Sitiveni Fe'ao Fehoko, collection of Te Papa; **169 top** Robin Slow; **bottom** Brian Flintoff; **170 top** Metua Tangiatua; **bottom** Brian Flintoff; **171 top** Owen Mapp, collection of Te Papa, photo Hanne

Eriksen Mapp; **bottom** Norman Heke, Te Papa;
172 Eric Lee-Johnson, collection of Te Papa; **173**
James McDonald, collection of Te Papa; **174** Eric
Lee-Johnson, collection of Te Papa; **177 bottom**
The Gisborne Herald; **180** Fairfax Media; **182**
Te Papa

WHALES AND DOLPHINS OF AOTEAROA NEW ZEALAND

189 Mark Jones; **192** Getty Images; **194**
Department of Conservation; **199** Rebecca
Hamner; **201 top and bottom** Steve Dawson,
from Liz Slooten & Steve Dawson, *Dolphins
Down Under*, Otago University Press, 2013;
203 bottom Kim Westerskov; **207 right**
Kim Westerskov; **210 top** Kim Westerskov;
212 top Kim Westerskov; **bottom** Rob Hunt;
213 top Carlos Olavarria; **second from top**
Keiko Sekiguchi; **second from bottom** Carlos
Olavarria; **bottom** Alexander Turnbull Library,
EP/1955/1031-F; **214** Kim Westerskov; **218
bottom** Kim Westerskov; **220 top** Tidegeist;
bottom Jochen Zaeschmar; **224** Alexander
Turnbull Library, 1/2-003190-F; **231** Keiko
Sekiguchi; **233** Keiko Sekiguchi; **234** Nan
Hauser; **237 bottom** Whanganui Regional
Museum; **247** Department of Conservation; **249**
Chris Riley, Eco Wanaka Adventures; **252 top**
Tim Cole/Otago University Auckland Islands
Expedition 2012; **bottom** Kim Westerskov;
253 Jean-Claude Stahl, Te Papa; **255** Simon
Walls, Department of Conservation; **256**
Kim Westerskov; **258** Smithsonian Institute
Archives, 2002-32264; **259 top** Dennis
Buurman; **bottom** Keiko Sekiguchi; **261**
collection of Te Papa, Gift of the Department
of Conservation (Auckland Conservancy),
1994; **263 top** Keiko Sekiguchi; **bottom** Otago
Museum, Dunedin, New Zealand; **265 top**
Hans Stoffregen, Department of Conservation;
bottom Kim Westerskov; **267 left and bottom
right** Kim Westerskov; **top right** Stephanie
Behrens; **268** Department of Conservation;
270 Kim Westerskov; **272** Department of
Conservation; **274** Kim Westerskov

INDEX